# Last Days of the Slocum Era

Volume One

Graham L Cox

Random Boats

Published in Australia by Random Boats

PO Box 1143, Buddina, QLD 4575, Australia

For more information or to book an event contact:

**yachtarion@hotmail.com**

**https://randomboats.blogspot.com**

Book design by Graham L Cox, with a bit of help from Atticus, an author's best friend

Cover design by Graham L Cox, with a little help from an author's best friend, Atticus.

ISBN - print: 978-0-646-89595-6
First Edition:  May 2024

# Contents

**Dedicated to the memory of Dr David Henry Lewis, 1917-2002. His friendship was one of the great rewards of my life.**

*We, the Navigators*

THE ANCIENT ART OF
LANDFINDING IN THE PACIFIC

*David Lewis*

To Graham Cox,
without your friendship & help the more southerly navigation now pending would scarcely have been possible
Best wishes & thanks
David Lewis

AUSTRALIAN NATIONAL UNIVERSITY PRESS
CANBERRA 1972

"ICEBIRD", Sydney, Oct 1972

## *Roadways*

*One road leads to London,*
*One road leads to Wales*
*My road leads me seawards*
*To the white dipping sails.*

*One road leads to the river,*
*As it goes singing slow;*
*My road leads to shipping,*
*Where the bronzed sailors go.*

*Leads me, lures me, calls me*
*To a salt green tossing sea;*
*A road without earth's road-dust*
*Is the right road for me.*

*A wet road heaving, shining,*
*And wild with seagulls' cries,*
*A mad salt sea-wind blowing*
*The salt spray in my eyes.*

*My road calls me, lures me*
*West, east, south, and north;*
*Most roads lead men homewards,*
*My road leads me forth.*

*To add more miles to the tally*
*Of grey miles left behind,*
*In quest of that one beauty*
*God put me here to find.*

John Masefield

**Boldness has genius, power and magic in it.**

Johann Wolfgang von Goethe.

(From a translation of Faust by Irish poet, John Anster.)

# Preface

Most of the boats covered in this work were designed and built in accordance with imperial measurements, and I have chosen to use those in the text. The exception is when the boat in question was built in Europe, and there I use the metric system. Depths of water are given in metres, which has been standard for a long time, although I still carry numerous old charts with depths recorded in fathoms. It's tricky, but I like it, as it keeps me in touch with the Slocum era. Heights are given in metres. Distances are always in nautical miles, as I am not much concerned how far away one place is from another overland.

I am an Australian, who grew up in South Africa with an English mother, and British English is my natural idiom. It harks to me of elegance and poetic virtue. For American readers, whose teeth may grate when they read my British English, I can only sympathise. I know what a jolt it gives me to read words like humanize.

Fair winds to you all.

Cover image: American-flagged schooner, *Ishmael*, at anchor off Dangar Island, Hawkesbury River, New South Wales, Australia, August 1979.

# Introduction

*Last Days of the Slocum Era, Volume One,* is written for people, like myself, with an unquenchable thirst for stories about cruising under sail. I am my own ideal reader, but I know there are many others. It spans the years between 1966 and 1980, and the companion book, *Last Days of the Slocum Era, Volume Two,* covers the period from 1980 to 2024. It is partly a sailing memoir, and partly a review of the changes that have taken place in the cruising world, both technical and cultural, during those 58 years.

Although it examines aspects of my personal life that had an impact on my cruising aspirations, they remain something of a backstory. It is mostly about the boats I met and sailed aboard, and their people. As that extraordinary ocean voyager, Pete Hill, said to me in 2016, not everybody is a boat nut. He and I both are, and I assume that the reader is too. You'll find everything in here, from ultralight sloops through to gaff and junk-rigged vessels, catamarans and trimarans, and even Pacific proas, which I consider to be the purest form of sailing vessel ever conceived.

Not that I'm much of a sailor, at least not by my standards. Since I first became interested in ocean cruising my mid-teens, I have met some wonderful sailing vagabonds, some famous, some unknown, and find it hard to measure up to them. I have tried though; it is all I have ever wanted to do. I'm not much good on the tools either, despite laying a few keels. Once again, I am comparing myself to some magnificent craftsmen, whose skills I would dearly love to possess.

Apart from reading, the only thing I have ever been good at, in a half-arsed sort of way, is slinging words together. At least that is how it seems to me. You'll have to be the judge of that. Well-crafted language is like beautiful music, seducing the mind, singing a paean to adventure and freedom. Stephen King, the best-selling American author, talks about the 'portable magic' of words.

Language is also integral to our worldview. It starts the moment we are born, with our mothers and other people whispering stories in our ears. Eventually, we adopt, or adapt, some of those stories for ourselves. From the beginning, I was one of those kids who were beguiled by stories. I have been told by some that I am a born romantic, while others claim that I never grew up.

My worldview was strongly influenced by the books I read as a teenager, those classic tales of sailing adventure from authors like Joshua Slocum, Harry Pidgeon, Peter Pye, Eric Hiscock, Miles Smeeton, John Guzzwell, Peter Tangvald, etc, through to the works of more recent ocean voyagers like David Lewis, Bernard Moitessier and Robin Lee Graham. I was also influenced by stories I heard from vagabond sailors on the International Jetty in Durban, South Africa, the city in which I grew up.

It was an era when voyaging sailors were more independent than they are today, allowing for a few exceptions. When they cast off and sailed across the horizon, that is exactly what they did. They *cast off*, embraced the solitude of the sea, and were rewarded with a freedom almost unimaginable to most sailors today. They didn't call home on their satellite phones, get real-time weather updates from the internet, or get tracked across the ocean by transponders. They were alone with the sea, and nobody knew where they were until they made their next landfall.

That is the world I experienced when I first went to sea in the early 1970s, and these influences have made me something of an iconoclast; although, like almost everybody in the $21^{st}$ century, I've made my compromises with technology.

I've tried to be as honest as possible, both factually and psychologically. I have been blessed, or perhaps cursed, with a powerful memory, but even so, it is a strange thing. It is a record filtered through perception, for one thing.

Some years ago, when I showed one of my early stories to my shipmate, Ros Dawes, about our 1976 voyage across the Coral Sea on the 32' steel sloop, *Ice Bird*, she said, 'That's not what happened.' Perhaps she should write her own story, as I'd love to read her account. Alternative histories are so fascinating. Ros is long since married to Barry Lewis, son of the famous voyager, Dr. David Lewis, who owned *Ice Bird,* and took the sloop to Antarctica in 1972.

Even when I have logbooks to fall back on, I sometimes encounter difficulty. For instance, I have a vivid memory that it was my shipmate, 'Kiwi' Jim Ferris, who first noticed *the thing* following our 60' gaff-rigged schooner, *Ishmael,* one dark night deep in the Southern Ocean, when we were sailing from Nelson, New Zealand, to Papeete, Tahiti. It was the night after we'd crossed the International Date Line on 25 April 1980, and I already felt a bit spooked about having two Fridays, one after the other. Using the nautical almanac to work out my sun sights made me uneasy. Was it really still 25 April? I knew it was, intellectually, but it felt wrong. You realise how arbitrary time and other human constructs are.

My logbook from that passage records that I was the one who spotted *the thing*, but I cannot shake the memory that Jim did, so have written it as remembered. Ultimately, it does not matter who saw it first. That fact does not alter the bones of a ridiculous story. So, bear with me, shipmates.

The title of this book is, of course, a provocation. It served the purpose, while writing, of focusing my mind on the primary themes of my thesis, i.e. the challenges, rewards and values of voyaging in the days before satellite navigation and communications revolutionised bluewater sailing, and the ways in which the new technology changed the practice and culture of what once seemed a timeless way of life.

As Marshal McLuhan said, many decades ago, the medium is the message. Not only has the way we cruise changed these days, so too has our perception of what we are doing. If the Slocum era started in 1895, when Joshua Slocum first poked *Spray's* bowsprit eastwards towards the open horizon, it was possibly buried by a tide of affluence and technology after 1985. But things are never that simple.

Ocean voyaging has always attracted unforgettable characters, it is the nature of the thing, and the 21$^{st}$ century continues to do so, even if there are a lot fewer such individuals than there used to be. One only has to think of people like Roger Taylor, with his diminutive, engineless, junk-rigged voyaging yachts, *Mingming* and *Mingming II*, veterans of the Arctic, or Alan Martienssen, in his engineless, junk-rigged Benford dory, *Zebedee*, or Hans Klaar, on his 70' Tuamotu pahi, *Ontong Java II*, and his friend, Kiana Weltzein, on her 40' Wharram Narai catamaran, *Mara Noka*, the latter two being admirably low-tech, organic boats, to realise that the spirit of the Slocum era is still extant, albeit the exception these days rather than the rule.

Even among the young, the kids who were born into the digital age, there are exceptions to my thesis, such as Garrett Tarter, a young man sailing the South Pacific alone in his Nor'Sea 27, *Hooligan*, inspired, like so many before him, by the first teenage circumnavigator, Robin Lee Graham. He has the cover of Robin's book, *Dove*, taped to the bulkhead alongside his chart table. Garrett was 19 when he left Hawaii in 2023 and sailed his inaugural, 30-day ocean passage to Tahiti. The trip threw up no exceptional weather, just the usual squalls, calms and re-enforced trade winds, but tested his resolve and skills. Ocean cruising, especially alone, is as much a psychological test as a physical one.

*Hooligan* can only be described as a simple boat. Garrett bought it for US$19,000, saved up while working during his high-school years, and his equipment is minimal by 21$^{st}$ century standards. He has a small diesel engine with limited fuel capacity, and carries his water supply in portable jugs. He also has a solar panel, an electric autopilot, a Monitor wind-driven self-steering gear, an AIS, which proved useful at times to alert him to shipping, which also provided GPS signals, and a VHF radio for short-distance communication.

He carries a selection of paper charts, an iPad that has a GPS connection plus electronic charts, and, until it failed en route to Tahiti, an iPhone that gave him limited satellite communications, mostly weather reports and brief welfare checks with his family. He could not afford to buy a liferaft, but had an EPIRB and a paddleboard. He acknowledged that the chances of survival, if *Hooligan* sank, would be marginal.

The VHF radio caused some temporary alarm when he accidentally piggybacked onto a conversation between two ships arranging a transfer of bulk fuel. The skipper of the supply ship kept repeating that he was bringing Garrett fuel, and asking for his coordinates. It freaked Garret out until he resolved the misunderstanding. What the incident also did was strengthen the realisation that had been growing in his mind for days, that he was alone on the ocean. Out there, alone on the deep-blue sea, nobody can hear you scream.

When his iPhone failed mid-passage, he not only lost his ability to receive weather reports and communicate with family, but his entertainment module as well, because he had not downloaded E-books and music to his iPad. But he had paper charts and print books, and an excellent understanding, for such a young sailor, of the tenets of good seamanship and passage planning. He carried on with little more than a shrug of annoyance, completing a textbook passage.

Garrett has a YouTube channel, *Garrett's Adventures,* perhaps a 21$^{st}$ century equivalent of the sailing books that inspired me as a boy. Besides excellent cinematography, his narration has focus and charm. His YouTube storytelling could be seen as a metaphor for the way that the values and methodologies of the past can be translated into the cyber culture of the digital age, informing modern practices without loss of relevance.

Most of us today are glad enough for a GPS position at 0300, especially in poor visibility and close to danger, even if we are not foolish enough to believe that electronic positions are infallible, let alone chart datum, and don't mention lightning strikes. There is nothing wrong with technology per se, it is more a question of using it appropriately, as E. F. Schumacher so elegantly argued in his 1973 seminal work, *small is beautiful.* The error that so many make is to assume that bigger, more expensive, or more complex solutions are better.

*Last Days of the Slocum Era: Volume One* explores the world of ocean cruising before the technological and cultural changes mentioned above occurred, while *Volume Two* illustrates their impact, both on the author and on the cruising community as a whole.

# Chapter One

# A Lightbulb Moment

I do not come from a sailing family. My father could not swim and nearly drowned in World War Two, when he was a South African prisoner of war aboard the Italian ship, *Sebastiano Venier*, torpedoed by a British submarine in the Mediterranean in 1941. The ship was flying the International Red Cross flag to declare that there were POWs aboard, but Winston Churchill ordered all Axis ships to be sunk regardless. Before the war, my father feared the sea; afterwards he both feared and hated it.

When he returned to Durban, where I was born on 12 April 1952, he concentrated on securing as much capital as he could. Luckily, his great passion for cars opened up opportunities, despite coming from an impoverished background with little formal education. When he was not working in his Local Government job, which he clung to for security, he slaved in his workshop fixing cars. He was thus able to buy a house, educate his four boys, and build a substantial investment portfolio.

As for me, I had no interest in the sea. I enjoyed a few books about sea adventures, like *Tintin and the Pirates*, being romantic by nature, although the political subtext in the Tintin stories escaped my attention. My ambition was to be a ballet dancer, or some sort of prancing thespian, though my father left me in no doubt that it would be over his dead body. I wasn't a happy boy.

Then, in late 1966, at the age of 14, I read a two-page article in a Sunday newspaper, with lots of photos, about a local yacht, *Sandefjord,* that had just returned from a 22-month circumnavigation. The article detailed how two Durban men in their mid-20s, Barry and Patrick Cullen, had found the ship half-sunk in the harbour and spent two years rebuilding it, before setting sail with four others on their grand adventure.

*Sandefjord sailing back into Durban Harbour in November 1966, at the end if its circumnavigation. Photo: Sandefjord archives.*

The article went on to say that *Sandefjord* was a genuine Norwegian rescue vessel, redningskiote number 28, launched in 1913 at Risor, 46' long, and weighing 50 tons, designed by the legendary Colin Archer. The boat had once been owned by a famous sailor, the Norwegian, Erling Tambs, who'd been forced to sell it in Cape Town when World War Two broke out. Under the command of Tambs, *Sandefjord* had pitchpoled in a hurricane in the North Atlantic in 1935, with the loss of one crew member, the mizzen mast, and a topside plank.

It was not uncommon for the local newspapers to publish articles about visiting yachts, though I'd never taken much notice before. Durban, in those days, was still among the legendary ports of call for voyaging yachts, along with places like Cape Town, Rio de Janeiro, Falmouth, Barbados, Papeete, Whangarei and Brisbane. Early voyagers were feted in many of these destinations, but they all noted the exceptional welcome they received in Durban, and it became an important staging post for subsequent world-cruising yachts.

What is perhaps more notable is that this welcome was still evident in the late 1960s. Every time a yacht arrived in port, illustrated articles appeared in local newspapers, and the International Jetty, where visiting yachts moored in the heart of the city, was crowded with onlookers, some of whom were keen to offer hospitality and assistance. In Durban, ocean voyagers were rock stars, at least among the general public.

In the Point Yacht Club (PYC), as I soon discovered, there was a small cohort of local sailors who resented visiting yachtsmen, claiming they were straining the facilities paid for by regular members. This small cohort, propping up the bar that overlooked the foreign yachts rafted at the International Jetty, loved nothing better than making disparaging remarks about the visitors at every opportunity. I suspect the real problem was that the visiting yachtsmen, who had crossed at least one ocean to get there, made these dissenters feel inadequate. The majority of PYC members were both interested in and hospitable towards foreign yachts.

The PYC itself offered visiting yacht crews a warm welcome, with free membership for the first month. The PYC had been founded in the late nineteenth century by working men and retired sailors who lived nearby at the Point, and had a long tradition of offering hospitality to visiting yachts, going back to the days of Joshua Slocum and Harry Pidgeon. It was justly one of the most famous yacht clubs in voyaging circles for many years. It is still welcoming, even though most visiting yachts prefer to stop at the relatively new port of Richards Bay these days, 87nm to the north, and often bypass Durban on their way south.

For reasons I don't remember, probably because of my romantic inclinations, I went down to the yacht harbour, or yacht mole as the locals called it (there was a long, concrete mole behind which the yachts sheltered from SW gales) and had a look at *Sandefjord* where it lay alongside the International Jetty.

*Sandefjord* looked so massive, with masts like tree trunks, that I found it impossible to believe the sea had once upended it. Bernard Moitessier, whom I would soon come to consider a source of profound wisdom, called *Sandefjord* a great big bull of a boat when he saw it in Cape Town in 1956, during his voyage on *Marie Therese II*.

*Sandefjord* was a carvel-planked timber ketch, and you could see every plank in its topsides. There was something delightfully organic about the boat, with its white hull and spars, and well-scrubbed, bare timber deck. Its husky, work-boat style was to become my favourite yacht aesthetic.

I later discovered that this jetty was famous in voyaging circles. The yacht mole was built in the late 1940s, and since then many well-known sailors had moored there; Bernard

Moitessier on *Marie Therese II*, Jean Gau on *Atom*, Marcel Bardiaux on *Les 4 Vents*, John Guzzwell on *Trekka*, David Lewis on *Rehu Moana*, Robin Lee Graham on *Dove*, to name just a few, and many others who sailed in from the far horizon, then departed again unheralded, apart from the memories they left behind.

Moored alongside *Sandefjord* was a freakish boat, so weird I could not imagine such a thing possible. It had two hulls. Besides the person I assumed to be the skipper, (he was male and I was naive, though in this case the assumption was correct) were two women and two babies. This boat had two of everything, it seemed to me.

*Rehu Moana being launched in 1963 in England. Photo: Tony Jennet.*

It was a 40' catamaran, I soon learned, called *Rehu Moana*, and was painted bright red, with strange black squiggles on the hull, like artistic graffiti, and pale blue decks. I discovered later these that squiggles were traditional Polynesian glyphs. As I gazed in wonder, a short man stepped off the catamaran's expansive decks and crossed the far more decorous deck of *Sandefjord* onto the wharf. I smiled at him.

'Hello,' he said, in a soft, English voice, 'I'm David Lewis. Would you like to help me carry a box down from the yacht club?'

To my astonishment, he told me he'd sailed *Rehu Moana* almost all the way around the world, though I didn't know enough to be particularly impressed when he said he'd gone through the Patagonian Channels rather than using the Panama Canal.

*Rehu Moana* was quite a reasonable size for ocean voyaging, but in my ignorance, I thought it looked small and fragile. Compared to *Sandefjord*, of course, it was. There were no hefty planks visible in its hull structure. But *Rehu Moana* was possibly the most seaworthy catamaran of its time, designed by a prominent English naval architect, Colin Mudie, in conjunction with the Prout brothers, who built the boat very strongly in cold-moulded timber. The Prout brothers later led the way in production-built cruising catamarans.

I could not know that this man would later play a prominent role in my life. Nor did I know he was famous, having participated in the inaugural Singlehanded Transatlantic Race in 1960, alongside Francis Chichester and Blondie Hasler, among others, and written a couple of books (he wrote 11 seafaring books in his lifetime, plus a few on cultural matters). After that day, I returned to talk with him several times, although he often seemed to be rushing off somewhere, and there was an atmosphere of bedlam on the catamaran.

Once, David drew my attention to a small yacht called *Vertue Carina*, moored just ahead of *Sandefjord*, saying it had just sailed into Durban from Hong Kong, and was a sister-ship to the boat he had raced across the Atlantic in 1960.

*Vertue Carina on fore and aft moorings in Durban yacht basin.*

'She'd be the perfect size for a young fellow like you,' he said, 'but you should consider putting a junk rig on it. I think it's the ideal rig for a singlehanded sailor. I raced against a junk-rigged Folkboat once, in 1960, called *Jester*. Blondie Hasler could handle the whole thing from inside his cabin. Never got his feet wet, unlike the rest of us. I tried junk rig on *Rehu Moana* initially, but we didn't have time to develop it properly. If I was going to sail alone again, I'd seriously consider it.'

'Did Blondie Hasler invent the junk rig?'

David laughed. 'No, it's a traditional rig. It has been used by Chinese traders on their junks for centuries. Blondie just came up with some clever ideas to simplify it. Mind you, Joshua Slocum used it long before Blondie.'

I looked askance at *Vertue Carina*. Moored just forward of the massive bowsprit of 50-ton *Sandefjord*, it looked like a toy. David read my expression. 'She may be small,' he said, 'but you can go anywhere in one of these Vertues. *Cardinal Vertue,* the boat I took across the Atlantic, recently sailed around Cape Horn with her new owner, an Australian fellow called Bill Nance.' I had no idea where Cape Horn was. I thought he was talking about Cape Town.

*Bill Nance at the helm of Cardinal Vertue.*
*Photo: unknown.*

I came to realise later that the Vertue is indeed an astonishing voyaging yacht. More Vertues have probably made ocean voyages than any other single class of yacht. Designed by Laurent Giles in 1936, and based on the lines of a pilot cutter, this 25' boat has a beam of 7' 2", a draft of 4' 6", and a displacement of 4.2 tons, though people often refer to it as a 5-tonner, referencing its Thames measurement. It has a 47% ballast ratio, slackish bilges, 384 sq ft of sail, and is known for remarkably fast daily runs on passage, plus it has an enviable record of surviving heavy weather.

Like all English boats of the type and era, its slack bilges and narrow beam cause it to roll a bit when sailing downwind. The fibreglass version produced later has 8" more beam. Nonetheless, there is something about the Vertue that took hold of people's imaginations, and it is said that more Vertues have crossed oceans than any other single class of yacht.

*Jester at the start of the 1964 OSTAR. Photo: Jester archives.*

Folkboats, like Blondie Hasler's *Jester*, are perhaps even more remarkable. At 25.2' and 2.12 tons, they are less than half the displacement of the Vertue, are much more lightly built, and have a lot less internal volume. Nonetheless, they have also made numerous ocean passages. In later life, David Lewis told me that the Stella, a Kim Holman Folkboat derivative, was, in his opinion, a better sea-boat than the Vertue. I am not sure that I agree. It may be faster in light weather, and with the wind forward of the beam, but I'd prefer the heft of a Vertue any day for ocean voyaging.

There was another boat nearby called *Suhaili,* a 32' carvel-planked, bermudan ketch that was on its way from India to England. I remember thinking that *Suhaili* looked like a miniature version of *Sandefjord*, and later realised that is exactly what it was. William Atkins based the design, which he called Eric, on a scaled-down redningskiote. *Suhaili* was a lot bigger than a Vertue, but still looked small to my inexperienced eye. The boat's skipper, Robin Knox-Johnston, would soon become a household name in yachting circles, but was then an unknown merchant seaman. I never met him, and *Suhaili* left for Cape Town, and on to England, a few weeks after *Sandefjord's* return.

At the time though, it didn't mean much to me. David sailed off, *Sandefjord* went out to a distant mooring in the harbour, and I went back to school. It was just another summer distraction, one of my infamous enthusiasms that came and went. The only enthusiasm that never waned was my love of ballet. I would have danced till I dropped if I had the chance. When I was dancing, the music seemed to enter my bones, I became a part of the score, and escaped from misery into a magical place where everything was beautiful. I had no intention of getting involved with boats. It seemed to me to be something that old people did.

I doubt I gave sailing another thought until 21 October 1967, when I picked up the evening newspaper after dinner. After sneering at the political headlines with my rebellious, anti-apartheid perspective, I turned the page. It was the picture that took my attention first, a young man standing on the deck of a small yacht, looking shyly at his toes. Large eyelashes shielded his gaze, and an outrageous swirl of bleached-blonde hair framed his face. The hair excited me as much as anything. At my school, hair had to be at least two fingers above the ears. We all longed to grow our hair long. It amuses me now, in 2024, that some boys *choose* to have their hair shaved two inches above their ears. Somehow, though, they contrive to make it look more artistic than the convict haircuts of my youth.

I scrutinised the article. Unfortunately, I lost the newspaper clipping later, so the following words are taken from memory, but I can vividly remember turning the page, and the totally unexpected reaction it provoked.

*American Teenager Sails the World Alone.*

*18-year-old Lee Graham sailed into Durban Harbour today aboard his 24-foot yacht,* Dove, *after a rough passage from Mauritius. He left California two and a half years ago in the small yacht his father bought for him, and will be the youngest person to sail around the world alone by the time he returns.*

I am not sure why the newspaper reported his first name as Lee, but that is what I knew him as for some time afterwards. When seeing those days through the eyes of my teenage self, I'll continue to call him Lee, reverting to his proper name otherwise.

Looking at the article, I felt the birth of the most outrageous idea. Alone... aboard my own little yacht... sailing the oceans of the world... free to go wherever I pleased... nobody to ask me awkward questions... nobody to tell me what I could or could not do...and if they did, I would just sail away again. You've heard the expression, *a light bulb going off in your head*. It was just like that.

*Dove departs San Pedro Harbour on 27 July 1965, bound for Honolulu. Photo: Lyle Graham. Courtesy Robin Lee Graham.*

As soon as I could, I rushed down to the International Jetty and gazed in awe at *Dove*. It had some sort of contraption on the stern, rotating idly in the breeze. I noticed some of the other yachts had similar setups, if more compact. I later recognised this as *Dove's* wind-driven self-steering gear. By 1967, most yachts carried these systems, but not all. Some still relied on twin staysails, or other sheet-to-tiller arrangements. There was even a little controversy about self-steering gears at the time. There were those who believed that hand-steering promoted better seamanship.

I was a bit unnerved to see how small and frail *Dove* looked compared to some of the other voyaging yachts rafted alongside the jetty. It looked like a plastic toy, with large windows and fragile fittings, but the sea-stained white hull and sun-faded green decks spoke of faraway places.

Undoubtedly, *Dove* made *Vertue Carina* look more substantial, but I reasoned that if Lee Graham could sail *Dove* half-way around the world, then obviously it was just as seaworthy as a Vertue. I had learned something since the previous year, for instance that Vertues are special, but obviously I had not learned enough. It was to cause me some trouble later.

*Robin Lee Graham sailing Dove out of Honolulu after his trip from Los Angeles, with Diamond Head in the background. Photo: Lyle Graham, Courtesy Robin Lee Graham.*

I later discovered that *Dove* was barely seaworthy. It was a Lapworth 24, designed for daysailing in semi-sheltered waters, plus the deck had a plywood core that was rotting at the gunwales, and the boat had already lost its mast twice at sea. Recently, in a severe storm off Madagascar, one of *Dove's* portholes had been stove in when it popped out of its frame, and the companionway doors cracked. The cockpit was way too large.

One interesting feature of *Dove* that I did not notice at the time was that the rig included a roller-reefing jib. Most boats didn't use roller-reefing sails in those days, and there was widespread distrust of them among voyagers. They had a reputation for mechanical failure, plus poorly-setting sails when reefed. In Robin Lee Graham's book, *Dove*, he makes no mention of problems with his roller-reefing gear, and today, with improved technology, they are in widespread use, though many of them seem unnecessarily complicated and expensive to me. Years later, I discovered one unit that was so robust and basic that even Slocum might have approved of it.

It would be some time before I discovered that Lee had lost faith in *Dove's* seaworthiness by this stage, and was plagued by loneliness. He had no desire to resume his voyage, and only did so because of external pressures. Adventure had turned to obligation, but there is no doubting his courage, and his determination to do the right thing by his supporters. Years later, *Cruising World* magazine jokingly referenced his voyage as an example of the *Anything is Possible* school of ocean voyaging.

Nonetheless, Robin Lee Graham's voyage met all the criteria for competent seamanship and independence that made the Slocum era so admirable. Even before he left Los Angeles, he had mastered the intricacies of celestial navigation. On his first passage, en route to Honolulu, he took a series of star sights one bright, moonlit night when the horizon was clearly visible, allowing him to pinpoint his latitude and longitude on the chart, using a cocked hat of intersecting position lines, telling his tape recorder afterwards, *that's kind of fun*. Later, he built two successful jury-rigs and made it into port unaided after being dismasted at sea. Not having a radio-transmitter or sat-phone, he just got on with it, like generations of ocean voyagers before him.

He also felt the deep awe that many days of solitude at sea can bring, something that even crusty old Slocum reported. After taking his star sights, Robin mused about the nature of the universe. In *The Boy Who Sailed Around the World,* he writes, *I thought how fascinating it was that the sun and the moon and the stars keep much better time than any man-made clock or watch. America has a master clock at the National Observatory in Washington, D.C. But this master clock has to be corrected occasionally by checking it against the master clock of the universe.*

*Robin Lee Graham was a competent navigator and resourceful seaman, in the best traditions of the Slocum era.*

In many ways, a voyage like this, especially in those days when you were truly alone at sea, is as much a journey into yourself as it is to a destination, and Robin's voyage was to profoundly affect both his personal values and his path in life.

While I was gazing in awe at *Dove*, Lee appeared, walking down the jetty. He glanced at me briefly, his long, bleached hair looking like it hadn't been combed for some time. My heart skipped a beat, and I remember feeling intensely self-conscious about my uncool clothes and convict haircut. My parents had firm ideas about what they considered appropriate dress, and it didn't include cut-off jeans and scruffy t-shirts. I was too shy to speak to him, and he stepped down onto the deck of an American ketch, *Karen Margrethe*, moored alongside the jetty, before clambering across the rafted boats and disappearing into *Dove's* cabin.

Although I was too shy to approach him, I took to following him around like a puppy whenever I could get away with it unnoticed. They didn't call it stalking in those days. I also spent as much time as possible on the jetty, trying to peer in through *Dove's* hatches and windows. The boat was a bit too far away for this to be satisfactory, being rafted outside two other yachts, so I peered at them instead.

From then on, when I wasn't on the jetty, I spent most of my free time scheming to get a deck under my feet, or with my head buried in books about ocean voyaging, having discovered a vast horde in the Durban Central Library, housed in the back of the City Hall beneath the museum. It was an impressive building, modelled after the Belfast Town Hall. I loved that library, with its high, sculptured ceilings, wide stairways, sweeping wooden banisters, deep silence and thousands of books. This vast collection of stories had always been an irresistible beacon. I miss the silence of that place, where you could almost hear a spider spinning its web, unlike modern libraries.

The first two books I borrowed were *On the Wind of a Dream,* by Victor Clark, and *Once is Enough,* by Miles Smeeton. I chose the latter because of its outrageous title. Once was most definitely not enough. What was the man thinking? Just because they had been pitchpoled off Cape Horn.

But the joke was on me, as it so often is with sneering teenagers, for Miles and Beryl Smeeton were among the most adventurous and accomplished ocean voyagers you could imagine. After their two attempts at doubling Cape Horn recounted in *Once is Enough*, they had continued cruising aboard *Tzu Hang*, their 46' carvel-planked ketch, and were still at it when I first read the book in 1967.

I discovered this to my chagrin when I brashly told some sailor in the Point Yacht Club that a mere pitchpole would not be enough to stop me, and was informed that it hadn't stopped the Smeetons *either* (my interlocutor was kind). He told me that *Tzu Hang* had visited Durban a few years earlier, during an eastabout circumnavigation of the world from England, via the Red Sea, East Africa, Japan, the Aleutian Island, and Panama. They were, I soon discovered from reading yachting magazines, about to embark on their third, and ultimately successful, rounding of Cape Horn, this time from east to west, against the prevailing winds, after which they sold the boat and retired ashore. They could have written a book about that last voyage called *Three Times is Finally Enough*, but called it instead, *Because the Horn is There*.

In 1973, a couple of years after the publication of that book, they produced their final sailing narrative, *The Sea Was Our Village*, which recounts *Tzu Hang's* first ocean voyage, not previously told, from England to British Columbia via the Panama Canal in 1951, when they were neophytes. This was exactly the sort of voyage I wanted to undertake, sailing through the tropical trade winds to distant islands, across an azure sea, sails billowing, a warm sun on my back. It is a delightfully-told story that captures the charm of voyaging in the mid-20$^{th}$ century, which some say was the high point of the bluewater game. That they started with no experience, taught themselves celestial navigation as they went along, and were able to rise to any occasion unassisted, added to the story's charm.

*Tzu Hang, some years after pitchpoling in the Southern Ocean, rerigged without the original bowsprit. Photo: Clio Smeeton.*

The other book I borrowed, *On the Wind of a Dream,* is a poetic story that swept me off my feet. It recounts Victor Clark's tropical trade wind circumnavigation in the 1950s aboard *Solace*, his 9-ton, 33' 8" carvel-planked ketch, designed and built by David Hillyard in England in 1929. Victor Clark made it all seem so normal, so doable, so joyous. He was a warm-hearted man, and his affectionate relationship with his West Indian crew, Stanley, inspired me as much as anything. As Victor Clarke put it: *Contact with different ways of life broadens the mind and breeds tolerance.*

When *Solace* was wrecked on Palmerston Atoll, in the South Pacific, he dragged the boat ashore and rebuilt it with the aid of the islanders, before sailing down to New Zealand for more permanent repairs. One of the anecdotes I loved was when he was departing Auckland, and one of his friends asked him if he had enough money. 'Yes,' he replied, 'I've got four dollars.' This shocked his friend, but he pointed out there was nothing you could spend money on at sea.

The Smeetons could have Cape Horn. I knew, even without experience, that I was not attracted to cold and stormy climates. I have always been a bit of a lizard. What I did not know was that bad weather at sea can be miserable at any latitude. Visiting the International Jetty on a warm, summer day, looking down at the voyaging yachts rafted together, and seeing the bonhomie of their crews, made it seem like unalloyed joy.

After reading these books, I returned to the jetty to gaze at *Dove* and the other visiting yachts with deeper understanding. I didn't realise, of course, that I was witnessing the last days of what I have come to call the Slocum era. Within a few decades, ocean voyaging would change so much that Slocum would not have recognised much of the gear aboard cruising boats, let alone deigned to go to sea on some of the yachts that ply the oceans today.

As late as the 1950s, little had changed since the days when Slocum left New England and pointed *Spray's* bowsprit east in 1895. *Spray* was engineless, as all sailing vessels were in those days, with oil lamps for lighting, and sails made from cotton and flax. Many yachts in the 1950s still had cotton and flax sails, although most had fitted modest inboard engines. If they had electrical lighting, it was very basic, and oil lamps were still common. They usually carried radio receivers for time signals and entertainment, run on dry-cell batteries, but almost none had two-way radio communications. Slocum would have understood those yachts, he could have stepped aboard one and sailed away.

Take, for example, Eric and Susan Hiscock's first circumnavigation (1952-5), in their 30', eight-ton, carvel-planked sloop, *Wanderer III,* which had a 4hp, hand-cranked Stuart Turner petrol engine that could move the boat at three knots in a flat calm, provided the sea was smooth, although its primary purpose was to charge the ship's battery. The boat had electric navigation and cabin lights to use when convenient, and a 12V socket just inside the companionway for a spotlight, but on long passages they still used oil lamps.

Their working sails were made from cotton, suitably mildew-proofed, rendering them a pleasing, if patchy, red-brown colour, and the storm sails were flax. There was no anchor winch, and ridiculously small halyard and sheet winches by today's standards. When the wind was astern, they could get the boat to self-steer with twin staysails poled out on either side, otherwise they hand-steered, three hours on and three hours off, for weeks at a time.

They were not young, tough adventurers, either, but appearances can be deceptive. They barely upgraded the boat for their second circumnavigation a few years later, replacing the 4hp motor with an 8hp unit of the same size, and buying a new mainsail and working jib made from Terylene (Dacron).

*Susan Hiscock at the helm of Wanderer III. Photo: Eric Hiscock.*

*Eric and Susan Hiscock in the cockpit of their beloved Wanderer III. Photo: Hiscock archives.*

After happily voyaging aboard *Wanderer III* for 17 years, they sold the boat and purchased a 49' steel ketch, *Wanderer IV*. Despite the extra comfort this new boat provided, and despite undertaking one and a half circumnavigations aboard, they were never really content with the boat. They always seemed to be fixing one of its complex systems, though they did note that having large water tanks, and therefore being able to shower frequently, meant they didn't suffer from the skin rashes and boils that plagued them on earlier voyages. Eventually, they returned to a 39', tiller-steered timber yacht, *Wanderer V*, for their final years. They gave one the distinct impression that *Wanderer III*, with its very basic arrangements, was the boat they loved the most.

Had I been smarter, I would have taken the measure of boats like *Solace* and *Wanderer III*, and set my sights on something similar, or at least a Vertue. But I had been influenced by *Dove*. To reinforce this impression, in *Once is Enough*, Miles Smeeton wrote about John Guzzwell, their crew for *Tzu Hang's* first, ill-fated, passage around Cape Horn, and his little *Trekka*, a mere 20' 6" yawl. Guzzwell had already crossed the Pacific Ocean, and was reported to have since pushed on westward across the Indian Ocean.

*Trekka outward bound from San Francisco towards Hilo, Hawaii, July 1955. Photo: courtesy John Guzzwell.*

I still remember the thrill when I found John Guzzwell's book, *Trekka Round the World,* on the library shelves in late November 1967. The excitement was so raw it hurt my throat. I have never tired of reading that book. As soon as I finished it the first time, I read it again. Guzzwell wrote with a simple joy that was mesmerising, and I soon bought my own copy. I once added up that I had read the book 15 times, and still dip into it. I can almost quote some passages from memory.

*It was really grand sailing, small puffy white clouds were all marching in order across a wonderfully blue sky, while down on the sea about us the waves flashed in the sunlight. Every once in a while Trekka caught a wave and surfed down the face of it, sometimes she slewed a little off course and I'd watch the tiller automatically correct her and bring her back again. So it went on, hour after hour, trekking across the ocean towards the distant islands.*

*...On 2 November, my twenty-eighth day out, we were only 120 miles from Hilo ...I woke up the next morning with the sunbeams dancing across the cabin. I stretched rather lazily and switched on the radio with my big toe. 'Let's see,' I thought, 'one hundred and twenty to go at noon yesterday. From the sound of her she's doing about three now. Yes, if it's clear, I should be able to just see the top of the mountain.' I promised myself that if the island was in sight, I'd have a can of peaches for breakfast. Sure enough it was peaches for breakfast, because when I looked out of the hatchway, there, right ahead, but still low down, was the 13,825-foot peak of Mauna Kea. I viewed it with mixed emotions. I was pleased to see it, yet the passage down from 'Frisco had been so enjoyable that I felt a little sorry that it was almost over.*

If there was ever a doubt that this new obsession was going to stick, something I think my parents were sincerely hoping for, *Trekka Round the World* put paid to that. John Guzzwell's enthusiasm, and the sunny tone of his writing, were enchanting.

# Chapter Two

# On the Wind of a Dream

One day when I arrived at the yacht basin, *Dove* was gone. I soon discovered that Lee hadn't left but had taken his boat to the Bayhead, a remote commercial area where the dry-dock, railway shunting yards, and various amateur boat-building projects were situated. Certain members of the Point Yacht Club had been rude to Lee, claiming he was just a rich American brat, and that anybody could do what he'd done if they had a rich daddy to pay the bills, which was wrong on both counts. They were also making disparaging remarks about him living 'in sin' with his girlfriend, Patti. It was the usual idiotic bar talk, fueled by alcohol, feelings of inadequacy, and envy, but it led to an unfortunate confrontation.

I had never been to the Bayhead, which was a long way from home, but the next Friday afternoon after school I rode there on my bicycle. I found *Dove* tied bows in to the shore, just past the floating dry-dock, with the fore-hatch propped open. I was standing on an old box, trying to peer into the boat, when Lee's sun-tanned arm appeared through the hatch, drawing it closed. *Oh well...*

There were some strange fibreglass things, long, narrow and box-like, on the foreshore, and I was looking at them, wondering what they could be, when an old Zulu watchman wandered over.

'Is clever, that boy,' he said. 'He make these things himself.'

'What are they?' I asked, indicating the fibreglass objects.

'He make new backside. Too much water big storm Madagascar. He pray God save him. He good, this boy. You pray God, you no get trouble. I pray this boy come safe America.'

He put a friendly hand on my shoulder and smiled. It was unusual in 1960s South Africa for a black person to be this familiar with a white person, even a boy. It may have been my youthful, friendly countenance, combined with the isolation of the area, but I suspect he was also encouraged by his interactions with Lee. Whatever the reason, I remember being delighted.

I went back to look at *Dove* as often as possible after that, but could only sneak down occasionally in the late afternoons after school, before my father came home from work. He kept me on a tight leash and was not to be trifled with; he could have taught a Yankee bucko mate a thing or two. Luckily, my mom usually colluded with me. I hoped I'd find Lee working on the foreshore, or on the deck, and that he might invite me aboard, but it never happened.

Nonetheless, the Bayhead became one of my favourite places. Besides the commercial shipping activity, there were several amateur boat-building projects in temporary sheds, mostly of ferro-cement construction, a popular building method at the time. With the cost of chicken wire and cement being quite low, it was possible to build large hulls for ridiculously small amounts of money. Of course, engines, rigging and sails cost the same as for any other yacht, and many of these hulls were never finished. Some of them could also more accurately be described as sculptures rather than engineered structures, but I was oblivious to such subtleties.

There was another Venus Trap at the Bayhead. The place was full of wrecks; old lifeboat hulls, launches, and even the odd sailing yacht that had seen better days. Some of these were subject to desultory efforts at restoration, and I began to dream of taking on one of them myself, as soon as I'd earned a little money. I'd climb up onto their rickety decks, if they hadn't rotted away completely, and squat there, mentally transforming their sagging timbers into solid wind-ships capable of carrying me to the West Indies and beyond.

Working my way through the cruising books in the library, I discovered that David Lewis had been a famous singlehanded sailor before setting off with his family on *Rehu Moana*, and soon knew the names of all the legendary singlehanded sailors; Joshua Slocum, Harry Pidgeon, Alain Gerbault, Jacques-Yves Le Toumelin, Marcel Bardiaux,

Bernard Moitessier, Ann Davidson, Val Howells, Francis Chichester, etc. Their names became icons, signposts on the path from misery to happiness.

I just assumed I was going to be a singlehanded sailor. Why is hard to explain, but I think it had something to do with low self-esteem, and the belief that I could never find someone to love me. Many of these legendary sailors had passed through Durban and were remembered around the yacht club. In 1967, there were a couple of elderly men who still remembered Slocum's 1897 visit, 70 years earlier, aboard *Spray*, when they had just been boys like me, as they told me with delight in their eyes. Quite a few more remembered Harry Pidgeon, the second solo circumnavigator, who arrived in Durban on 23 December 1923 aboard *Islander*.

Slocum was the hard-bitten professional seaman who'd come up through the hawsepipe and was prepared to defend his position. *Spray* was a rough old thing, an oyster smack he'd rebuilt from a wreck, with built-up topsides and concrete in the bilge for ballast. Slocum had fallen on hard times, and this voyage was his way of fighting back. He was successful, but his face showed his pain. In 1909, a decade after his circumnavigation, with *Spray* looking much the worse for wear, possibly due to poverty and mental health issues, he disappeared at sea. His dry, witty book, *Sailing Alone Around the World*, became one of the great classics of sea literature.

*Joshua Slocum on the deck of* Spray*, circa 1908. Photo: Slocum archives.*

*Islander*, by way of contrast, was designed and built as a yacht, although it was almost as basic as *Spray* in fitout. It had a distinctive, hard-chine hull, and a sweet sheer. Pidgeon was a generous man, with a pleasant, open face. He had a Quaker's willingness to accept people on equal terms, and managed to glide through life with ease. Before his circumnavigation, he led a peripatetic life, collecting specimens in Alaska for museums, drifting down the Mississippi River on a raft, and establishing a reputation as a fine photographer. He went on to complete a second circumnavigation aboard *Islander* and live to a ripe old age.

*Harry Pidgeon at the helm of Islander in San Pedro Harbour, Los Angeles, at the end of his second circumnavigation, December 1941. This image is taken from a newspaper clipping I found, among other memorabilia, in my autographed first edition of his book, Around the World Single-Handed.*

I find it interesting to compare these first two circumnavigators and their boats. One feature of the Slocum era, still evident when I was a boy, is that every boat that came into

Durban was as different as their skippers. They were mostly timber yachts, often old or home-built, providing me with a smorgasbord of delights. This colourful individuality has largely been lost now to mass production and cultural homogeneity. There's even a McDonalds in Tibet.

Slocum was not only the first person to circumnavigate the world alone, but was also the first Westerner to make a long bluewater voyage in a small boat fitted with junk rig. After the loss of his barque, *Aquidneck,* on a sandbar in Brazil in 1886, he built a 35' sampan out of the wreckage, fitted the boat with three masts rigged with junk sails, or as he called it, Chinese lug sails, and in 1888, with his wife and two of his sons, safely completed a 5000-mile voyage back to the USA.

He called junk rig *the most convenient boat rig in the whole world*. By the time I read Slocum's book, *The Voyage of the Liberdade*, I had already been alerted to the attractions of junk rig by David Lewis. As David suggested, it seemed to me like the ultimate rig for single or short-handed ocean cruising. I determined then that I was going to fit junk rig to my boat. I did, too, but I blush to think of how long it took.

I also liked the look of gaff rig, it seemed more robust and less complicated than the bermudan rigs I examined at the jetty. Besides *Sandefjord*, I'd taken a fancy to *Moonraker of Fowey*, Peter and Anne Pye's gaff-rigged cutter.

*Moonraker of Fowey. Photo: Peter Pye archives.*

Peter Pye was one of my favourite authors, although he was not a singlehander. Despite being a medical doctor, he tied his pants up with string, and wrote with an affectionate, unpretentious style. In 1931, along with his wife, Anne, he had purchased, for next-to-nothing, it seemed to me, an ancient Cornish fishing smack, 29' long, with a beam of 9' 8" and a draft of 6', then had it converted into a rudimentary gaff-rigged cutter called *Moonraker*, on which they cruised widely. Peter said the boat looked like a box and sailed like a witch.

*My sartorial guide, Dr Peter Pye, in Moonraker's cockpit. Photo: Peter Pye archives.*

*Moonraker* was a tired, leaky boat, with none of the shiny, expensive equipment advertised in yachting magazines. There were no electrics, a 6hp petrol motor that gave the ship, at best, 2.5 knots of speed in a calm, and no winches apart from an ancient two-speed anchor winch. The ballast, all internal and secured beneath the screwed-down cabin sole, consisted of old cannonballs, I kid you not, plus cogwheels and other stuff cast off from the early days of the Industrial Revolution. At times their voyages were a hard, wet, dangerous slog, but they joked about it. Anne once said that they sailed with *one hand in God's pocket*, and I came to consider this couple the epitome of *cool*.

I also admired Peter Tangvald. In his book, *Sea Gypsy*, he describes buying an old, carvel-planked cutter in England, the 32' *Dorothea*, designed by Harrison Butler, then removing the cockpit, engine, electrics, toilet and all seacocks, before sailing it around the world. He made a passionate case that this approach created a more seaworthy boat and led to safer cruising. With a silver earring in his left ear, he was a memorable character with iconoclastic views, who influenced me deeply.

*Peter Tangvald on the deck of Dorothea. Photo: Tangvald archives.*

He not only inspired me, but such a luminary as Larry Pardey, who built two yachts, *Seraffyn* and *Taleisin*, both without engines or electrical systems, the latter without a cockpit. Aboard these boats, Larry, with his wife, Lin, completed two extended circumnavigations, sailing in and out of hundreds of ports and anchorages. The Pardeys first set sail in the 1970s, when sextant navigation prevailed, but they chose to continue with physical navigation systems in the decades that followed, keeping the old traditions alive in more ways than one.

Peter Tangvald lost *Dorothea* not long after his circumnavigation, when the boat struck an unidentified object and sank en-route from French Guiana to Barbados. He rowed the last 50 miles in fresh trade wind conditions in his 7' plywood dinghy, a remarkable feat. Undeterred, for the next 35 years he voyaged on in his new boat, *L'Artemis de Pytheas*, a 50' engineless centre-boarder he built in French Guiana from local timbers.

He lost his life and that of his young daughter, Carmen, when that boat was wrecked one night on a reef in Bonaire in 1991. His 14-year-old son, Thomas, who'd already witnessed the deaths of both his mother and later his stepmother in tragedies at sea, survived, but was lost at sea on his own vessel off Brazil in 2014. There seems to have been some darkness at the heart of Peter Tangvald's story, and I learned later that he was a strong character who dominated those around him. He married seven times, two of those wives died at sea aboard his vessels, and his first two children also perished at sea.

*Le Artemis de Pytheas.*

But in 1967 I was not aware of any darkness, just raw excitement. I'd found, I thought, the ultimate way to give society the middle finger. After a summer of visiting the International Jetty whenever I could, and reading the thrilling books I found in the library, I developed a raging dose of sea fever. Some people call the summer of 1967 *the summer of love*. That was the northern hemisphere summer, of course, and it was a different sort of love, but I suspect there was a connection. There was definitely something in the air.

Oh yes, I was going sailing. Like John Guzzwell and Lee Graham, I would buy or build the smallest, cheapest, yacht in which it was possible to squeeze two narrow bunks, one on each side, so there would always be somewhere comfortable to wedge myself, equip it with a couple of kerosene lamps, a couple of buckets, and a single-burner Primus stove. I'd rip out the cockpit and deck it over, put a big windvane on the back, and then I would sail out of Durban Harbour *forever*. I'd work for three months a year, top up with beans and rice, and take off again. Ahead, I believed, lay the islands of paradise.

Robin Lee Graham sailed from Durban in April 1968, a few days before my 16th birthday. In what was perhaps a self-effacing gesture, he'd covered the brass letters of *Dove's* nameplate with white paint, so that you had to get right up alongside before you could read it. Perhaps he realised, with the imminent publication of the first *National Geographic* article about his voyage, that *Dove* would soon become famous.

But I knew I would recognise *Dove* anywhere, especially with its recent modifications. On Wilson's Slipway near the yacht basin, I had watched Lee, in a cloud of fibreglass dust and resin fumes, cut off the cockpit and replace it with a raised, flush deck. The fibreglass objects I had seen on the harbour wall at the Bayhead became *Dove's* new poop-deck coamings. I was too shy to talk to him, and he did not seem to notice me lurking in the shadows of the boatshed wall.

The last time I saw *Dove*, the boat was moored back at the International Jetty. I had a canoe by then that I could paddle around the yacht basin, when I could persuade my eldest brother, Bill, to take it to the harbour for me, strapped on the roof of his car. I paddled alongside *Dove* and gazed longingly at the boat, willing Lee to come out of the cabin, but there was no sign of life. The weather was inclement and *Dove's* hatch was closed.

I learned later that he'd actually left port a day or so earlier, but had to turn back because of headwinds. As he wrote in his second *National Geographic* article, *if the wind is forward of the beam, and 15 knots or more, I say forget it*. As I would discover myself later, beating to windward at sea in a small boat is the definition of being on a hiding to nothing. *Dove* left again a couple of days later.

When Lee finally sailed, it was late to be rounding Cape Agulhas, but he had been in no hurry to get back to sea, and the boat needed extensive repairs. He also married his girlfriend, Patti, in Durban. Those days were their honeymoon (occasionally interrupted by boys standing on boxes trying to peer through the forehatch). There was a short newspaper report a few weeks later saying that *Dove* was missing, but then he was found sheltering from a storm in an isolated bay close to Cape Agulhas. After that there were no more reports.

Another voyager who arrived in Durban in 1967 was Tom Corkill, a carefree, happy-go-lucky, 25-year-old New Zealander, sailing a pretty, 25' trimaran, *Clipper,* designed by Hedley Nicol. I was still too shy to approach adults, but after bumping into him in the Point Yacht Club car-park, when he asked me if I knew where he could buy some glue, I spent many hours talking to him, though he never invited me aboard *Clipper* and I was too shy to ask. His stories reinforced my ideas.

'You don't need a big boat,' he said. 'Go for the smallest, cheapest boat you can find. The smaller the boat, the bigger the adventure, and the more freedom you'll have.' He'd sailed *Clipper* from Brisbane, Australia, in early 1966, with no engine, lifelines, self-steering gear, electrical system or sextant, I seem to remember him saying, with few charts and even less money. I don't know how he located some of the more remote landfalls he made.

*Tom Corkill on Clipper in Singapore, January 1967. Photo: courtesy Alan Lucas.*

He was in Singapore on 10 January 1967, according to *Pacific Islands Monthly*, and then made a 62-day passage to Port Louis, Mauritius, before departing for Durban on 28 May. A 20-day passage brought him to Durban on 17 June. He therefore sailed through the volatile cyclone season in the tropical latitudes of the Indian Ocean on his way to Mauritius, and completed the daunting passage south of Madagascar to Durban in winter.

The undersides of the wing decks were painted lurid orange. When I asked Tom why, he grinned ruefully and said the designer, Hedley Nicol, had been uneasy with him taking this boat, intended for coastal sailing, across the Indian Ocean. What I didn't know, but Tom probably did by then, was that Hedley Nicol had left Brisbane in August 1966, bound east across the Pacific aboard his trimaran, *Privateer*, a larger version of *Clipper*, and had not been heard from since.

*Clipper sailing off Mauritius, May 1967. Photo: courtesy Alan Lucas.*

When Tom first arrived in Durban, all he had to eat were potatoes, but he was a slim, good-looking young man, with a dark fringe of hair falling across passionate brown eyes. With his boyish charm and imagination, he was soon everybody's darling. There were ten girls who wanted to sail with him when he left, but he chose solitude.

He set sail from Durban several weeks before Lee. In March 1968, an article he'd written appeared in *South African Yachting*, extolling the virtue of multihulls, especially 'advanced designs' like *Clipper*. Peter Pye was my hero, and his simple, rugged *Moonraker* still thrilled me, but multihulls seemed as unorthodox to me, in their own way, as old fishing boats, and in the early days were just as simple.

*The International Jetty, 24 October 1967. Clipper is on the left side of the jetty, nearest the land, while Dove is in the furthermost row of yacht on the right-hand side, third boat out. Photo: local newspaper clipping.*

A few days after reading Tom's article, the evening newspaper carried the story of how *Clipper* had capsized 200 miles NW of Cape Town in the Atlantic Ocean, on the way to Rio de Janeiro. Tom was safe but *Clipper* had been abandoned. I remember being incensed with the captain of the freighter for not rescuing *Clipper* as well. The thought of the boat just drifting around out there distracted me so much I could hardly sleep, and I spent my days at school in a greater daze than usual. Why, anybody could just go and grab it.

Another article appeared by Tom the following month, written on board the freighter that rescued him. He blamed himself for the disaster, not the boat. He'd over-loaded it, he wrote, not sealed the hatches properly, which probably let water into the outer hulls, and in the excitement of departure he hadn't rested enough, so that when heavy weather struck, he was exhausted and unable to tend to his craft. In moderate weather, he could get the little trimaran to self-steer with sheet to tiller, not having a wind-driven self-steering gear, but when it was too light, or, paradoxically, too windy, he had to hand-steer.

Unable to stay awake any longer, Tom took all sail down and let *Clipper* drift with the seas, which soon reached a height of about 30’. He was woken by a loud crash and the marrow-curdling roar of water flooding into the boat. Finding himself on the cabin roof, and realising that *Clipper* had capsized, he took a deep breath and swam out.

Clad only in a pair of shorts, clinging to the rudder, he was repeatedly swept off the slimy bottom into the sea and had to clamber back up. The cold began to take its toll and he resigned himself to death. He was determined, however, to hang on until the morning, to see one last sunrise. As dawn broke, a freighter steamed slowly past and sighted the orange under-body of the trimaran. Tom had been in the water for 18 hours. There was no chance of saving *Clipper*. In the heavy seas that were running, it was difficult enough to swim the few yards to the freighter and scramble up the cargo net they slung over the side. It took several days before he could walk unaided.

The punch-line was that he vowed his next boat would be another multihull. Properly prepared, he insisted, they were the safest boats in the world. I believed him, I believed everything ocean voyagers said in those days, even if they contradicted each other, but a little doubt niggled at me from then on.

I fell into the dream unexpectedly, but the seeds of enchantment were always there. I remember searching for beauty from a young age, declaiming over the petals of exquisite flowers, a bird, a kitten, or a sunset, and feeling a deep restlessness, a desire to escape ordinary existence. As a small child, I was a bolter, and had to be watched carefully in a crowd. I was always dreaming of Shangri-La. I suppose there are romantics born in every generation, lost to normative expectations and commercial enterprise, enchanted by beauty and stories of adventure. It was quite simple, really. The sea called and I answered.

# Chapter Three

# Summer Swallows

The harbour seemed empty without *Dove,* but there were always a few voyagers that stayed for the winter, to work and refit their boats. One yacht that wintered-over that year was *Mjojo of Lamu*, the most exotic boat that called into Durban during my years there. Built over a period of nine months on a beach on the island of Lamu, Kenya, by local boatbuilders using hand tools, *Mjojo* looked, to my eye, like an Arab dhow.

To some extent it was, but the project was the brainchild of a young English architect, Rod Pickering, and the hull-form was based on a model he'd made, influenced by English working boats. It was about 42' on deck, with a 15' bowsprit, a beam of 13', a draft of 6' 6", and a displacement of 24 tons or so, internally ballasted with pigs of iron in the bilges, like traditional working vessels. It had a boomless gaff-cutter rig, like a Thames Bawley, and was engineless, with no electricity installed.

*Mjojo's* dimensions were all a bit approximate, since the boat had been built more by eye than from any calculations. The local builders could not read plans, and even the model, whose form varied from their usual craft, raised some difficulties in interpretation. The construction was pure dhow. The carvel-planked hull was treenail fastened, and the vessel's topsides were oiled, with elaborate traditional carvings on the bulwarks and cutwaters. Apart from some modern materials in the rig, there was little to differentiate *Mjojo*, and the way the ship's complement crossed oceans, from *Spray* and Captain Slocum.

*Mjojo* oozed romance, exuding a tropical fragrance I could smell while standing on the jetty. I spent many days sitting on the pilings, looking at the boat, never tiring of its exotic looks that spoke of faraway places.

*Mjojo moored alongside the International Jetty. Photo: from local newspaper clipping.*

*Mjojo's* regular crew consisted of Rod and his wife, Dianne, their daughter, Jojo, Rod's friends from university days, Colin Frank and David Mitchell, who were fellow architects, and Lulu, Rod and Di's second child, who was born in Durban. Once launched, *Mjojo* had been sailed to Durban via the Seychelles and the islands of the Mozambique Channel. After staying for a year, they voyaged onwards to Cape Town, and then Rio de Janeiro via St Helena, before making the long haul up to England via the Azores Islands. England proved too cold and miserable after years in the tropics, so they retreated to Spain and then into the Mediterranean.

Eventually the boat was sold, as Di wanted to stay in one place and give the kids a shore-based life, while Rod did not. He went off to sea on various boats, returning to visit at irregular intervals. Some years later, after setting sail on a rickety, junk-rigged, 27' Wharram catamaran, he was never heard from again. It is presumed he was lost at sea, though some enigmatic messages from third parties over the years left Di wondering.

For many years, Di ran a charming little hotel, Los Castanos, in Cartajima, Spain, while her eldest daughter, Jojo, also became a sailor, and cruised from England to the Caribbean and back with her partner and teenage sons, in the gaff-rigged cutter, *Island Swift*. *Mjojo* ended up in a canal in Holland, where it languished for some years before being restored and returning to the Mediterranean. Di wrote an evocative book about *Mjojo*, titled

*The Ocean Voyager and Me,* and has also released a film of the voyage, *Mjojo Wa Lamu,* available on YouTube.

*Mjojo sailing on Durban Harbour. Photo: courtesy Di Beach.*

Another visiting yacht that stayed in Durban in 1968, although not by choice, was *Ohra*, a 25' 6" carvel-planked cutter from Lae, Papua New Guinea, where it had been built in 1963. Skippered by Laurie Harder, it had been in Port Louis, Mauritius, when *Dove* limped in from Cocos-Keeling atoll with a broken mast. In the same storm that damaged *Dove* off the coast of Madagascar, *Ohra* pitchpoled, losing the rudder. The skipper and his crew carried on under jury-rig, but when NE of Durban, *Ohra* was damaged further by a Japanese freighter that came alongside, trying to render assistance, so Laurie and his crew got off their yacht onto the freighter.

By the time they were put ashore in Durban, *Ohra* had washed up on the beach and been salvaged by cruising friends, including Robin Lee Graham and Tom Corkill, plus concerned locals. Nobody, at that stage, knew what had become of Laurie and his crew. When he arrived, dazed and despairing, at the Point Yacht Club, he found his boat propped up in the carpark.

Laurie spent a year rebuilding *Ohra* before resuming his voyage. The boat eventually completed a successful circumnavigation, despite another capsize en route to Cape Town. Just before it was relaunched, I remember standing in front of the bows, with my hand on the stem, imagining that soon *Ohra* would be rushing along through the South Atlantic trade winds to St Helena Island and the Caribbean. It gave me a rush.

Even though it is debatable whether *Ohra's* crew needed rescuing, the question remains what would have happened if they had not made it into Durban unaided, if perhaps they had been swept past by the current, something that has happened to a number of fully-found yachts. Like most cruisers in the 1960s, they carried no radio transmitter to call for assistance. It was also rare, in those days, to find yachts carrying inflatable liferafts. Bill Tilman, that extraordinary high-latitude sailor and mountaineer, famously said he thought carrying a liferaft indicated a lack of faith in your ship. Sailors like Tilman did not expect to be rescued if their yachts foundered. They would just disappear, or 'drown like gentlemen', to quote Blondie Hasler. Some did, including Joshua Slocum a decade after his world voyage.

It must be said, though, that most voyagers arrive at their destination. Ocean voyaging is a lot safer than many human endeavours, including driving your car down a freeway, something people do every day without thinking. That's probably one of the reasons why driving is so dangerous. Ocean voyaging was always safe, long before yachts became swamped with technological aids to navigation and safety. Maybe it was even safer then, on average, because the ship's company paid attention to staying afloat and getting themselves home.

In June 1968, the third Singlehanded Transatlantic Race started from Plymouth, England. It was now called the Observer Singlehanded Transatlantic Race (OSTAR) and there was a South African entrant in this edition, *Voortrekker*, a light-displacement, cold-moulded, 49' timber sloop, designed by Van der Stadt and skippered by Bruce Dalling, who'd sailed *Vertue Carina* into Durban in 1966. With a South African entrant, Durban newspapers provided extensive coverage of the race and the history of the event.

There were photos and stories about David Lewis and his boats, *Cardinal Vertue* and *Rehu Moana*, although he was not taking part this year. Having reluctantly sold *Rehu Moana* because he needed a vessel with reliable auxiliary power, he was heading back into the Pacific on an old Scottish ketch, *Isbjorn*. The plan was to continue his research into traditional Polynesian navigation techniques, begun during the circumnavigation on *Rehu Moana*.

One of the stories that interested me most was a profile of the junk-rigged Folkboat, *Jester,* developed by Blondie Hasler, a wartime hero and inventive genius. I remembered David talking about this boat. *Jester* was called the emblem of the OSTAR, being the only yacht to have participated in all three races. Already, the 1960 race had attained legendary status.

*Jester at the start of the 1992 OSTAR. Mike was 75, and competing in his ninth race. Photo: Rick Tomlinson, courtesy Nature PL.*

There was a detailed examination of what made *Jester* special, such as the ability to handle the rig entirely from the safety of the circular central hatch, and the rig's low-tech, low-stress features. Blondie was quoted as saying the rig was 'fully automatic'. There was no need to get your feet wet. The sail was controlled by just four adjustable lines, all of which led back to the central hatch; the halyard to pull it up, the sheet to pull it in, a yard-hauling parrel to hold the yard into the mast, and a luff-hauling parrel to give some tension to the luff, as the sail sat alongside the mast rather than pressing against its back, as in conventional rigs.

To reef or furl the sail, you just eased the halyard, and a batten or two dropped down. This slacked off the other three lines, which were then re-tensioned, and off you went. No getting out on deck to tie in reefs, change sails, or pole out headsails when sailing downwind. It appealed to my lazy and somewhat timid nature. Staying in the cabin seemed both more comfortable and safer. Of course, you still had the issue of staying afloat, something that was to give me some anxiety in the future. I thought *Jester* looked odd, but peculiarly exciting, and the unorthodox wisdom behind the boat*'s* concept appealed to my teenage mind.

Blondie Hasler was no longer sailing *Jester*, which had been taken over by a friend of his, Michael Richey, who went on to cement the boat's reputation. There cannot be many yachts with such a high profile that had a smaller material footprint, except *Trekka,* perhaps. *Jester* never had an engine or electrical system aboard. Later, Mike made a small concession to modernity by using a hand-held GPS unit for obtaining position fixes, operated by dry-cell-batteries. He then transferred *Jester's* position to his paper charts.

After Mike brought the GPS unit aboard, he stopped taking sextant sights during his Atlantic crossings, which shocked people, since navigation had always been one of his primary interests. His response, in an interview published by *Cruising World* magazine in January 1997, was to say, *What's the point? I don't need the practice.* He said it was not a navigational problem, but a cultural one. The world had changed and there was no way back. *That said,* he added, *I do think that hand-held GPSs fit in with* Jester's *concept of simplicity very well.*

Mike's minimalist existence aboard *Jester* gave him a great deal of joy. Perhaps it appealed to the side of his nature that had tempted him to become a monk in his youth. In the conclusion of that *Cruising World* article, he suggested that perhaps he should have become one, as he found it an attractive way of life. Then he smiled, saying, *I doubt I would have succeeded anyway. You need an awful lot of curbing, you know.* The sea became Mike's monastery, or his mana, as the Polynesians would put it. I found his charisma infectious, and his approach to ocean voyaging became a template for my own aspirations.

He was a brilliant navigator, trained by the Royal Navy in World War Two. He first achieved recognition in the yachting world by navigating racing yachts after the war. On several occasions, such as the 1959 Fastnet Race aboard *Anitra*, his star fixes in difficult conditions made the difference between winning and losing. At the request of the Royal Geographic Society, he set up the Royal Institute of Navigation, serving as its director between 1947-82. He also founded and edited the *Journal of Navigation*.

*Mike Richey in Jester's snug cabin. Photo: Jester archives.*

When I mentioned *Jester* to sailors at the Point Yacht Club, I met almost universal scorn. The boat was uglier than a baboon's arse, one told me, and the unstayed mast would snap like a rotten carrot. Most said the junk rig was too inefficient. Considering myself something of an expert by this stage, having read all about it, I was tempted to ask how a highly-decorated, retired Royal Marine like Blondie Hasler, who had specialised in small boat operations in World War Two, could have got it so wrong, but held my tongue, not wanting their disapproval.

I could also have pointed out that *Jester* came second in the 1960 race, with an elapsed time of 48 days. In the 1964 OSTAR, Hasler had shaved 10 days off that time, even though he came in fifth. On both return trips to England, *Jester* had made very fast downwind passages. The boat had suffered no damage, and that unstayed mast was still standing, like a middle finger to the naysayers.

I decided to wait, thinking that this time *Jester* might win and silence the critics. But times had changed. *Jester* came in last, with an elapsed time of 57 days, 10 hours. Mostly, this was because of the large, fast yachts now competing, plus the fact that Mike Richey took the far southern route, extending the distance from 3000 to 5000nm. But he wasn't really racing; he did it for the experience, to meditate in the vastness of the ocean, and to practice navigation. He admitted that in those early days he found navigation more interesting than sailing. He took up sailing in the first place because he did not want to put down his sextant after the war.

Despite *Jester's* performance in the race, my enthusiasm for junk rig remained, especially after reading that Mike arrived with a nice suntan, and said his only complaint was not having enough wine and books aboard. That sounded like my sort of sailing. Junk rig, I eventually realised, is a brilliant option for cruising yachts, but don't choose it for ocean racing if you want to win, especially if racing to windward.

When interviewed after the race, Mike said he couldn't wait to get back out there. He did, too, entering every OSTAR between 1968 and 1996. In the years that he crossed the finishing line, he also cruised back home alone. In all, he made 13 singlehanded passages across the North Atlantic Ocean in *Jester*, plus several shorter voyages to the Azores Archipelago and elsewhere. The original *Jester* was lost in a storm in the 1988 OSTAR, but the Jester Trust built him a new one. Mike said he knew it was a new boat, but in his mind it was always the same old *Jester*.

In 1997, sailing back from America, he became, at 80, the oldest person to make a solo Atlantic crossing under sail. He set out again in the 2000 OSTAR, aged 83, having failed to find a new skipper to take over *Jester*, but retired to the Azores Islands, and then cruised home. It became his last passage. *Jester's* simplicity, especially its junk rig, played a major role in his achievements, and his ability to carry on for so long.

He once mentioned that he did not want to die at sea. 'Give me clean white sheets and a priest,' he said, referring to his battle to survive the ultimate storm that engulfed Jester in August 1986. He achieved his wish and died at home in 2009, aged 92.

Bruce Dalling came in second on *Voortrekker* in the 1968 OSTAR, with an elapsed time of 26 days, 13 hours. He was 17 hours behind the winner, Geoffrey Williams, on the 57', Robert Clark-designed, *Sir Thomas Lipton*. Controversially, Williams pioneered weather routing in this race, getting advice via SSB radio, using a secret code, from a shore-based adviser running a weather-prediction program on a mainframe computer. It allowed him to skirt around a storm that stopped Dalling for several days. Weather routing was subsequently banned in the OSTAR, but has since become standard practice in long-distance offshore racing, and for many voyaging sailors. History may have vindicated Williams, who was vilified by some in 1968, but it was certainly a shot over the bows for the Slocum era.

One day in December 1968, I was talking to a man on the International Jetty. We were looking at a newly-arrived voyager, and I commented knowledgeably about this and that, sounding impressive even to myself. I had been reading a lot of books.

When I drew a breath, he asked casually, 'Have you seen the latest *National Geographic*? It has an article in it about that young man who's sailing around the world singlehanded.'

'What was his name?' I gasped.

'Robin something, I think. He came through here last year.'

'You mean Lee Graham on *Dove*? His name wasn't Robin.'

'Oh, I think that's what the article said: *A Teen-ager Sails the World Alone*, by Robin Lee Graham. I was just looking at it in the Royal Natal Yacht Club.'

My lips curled in scorn. *Stupid old fart. Imagine getting Lee's name wrong.* But there was no time to argue the point. I had to get that magazine before someone else did. They might steal it. At least this guy hadn't. I gave him a forgiving smile.

'See you later,' I said, and ran off through the car park, across the railway lines, across the four-lane highway, dodging cars, no time to wait for a gap, and into the plush foyer of the club.

I thundered up the polished wooden stairs to the reading room, oblivious to the disapproving glare of the club secretary. He was one of those men whose shirts never look creased, and whose ties always hang straight. He wore large, black-framed, thick-lensed glasses, through which his eyes peered at you like goldfish. The RNYC was not a place where members hurried.

Maybe the Indian waiters, carrying their silver trays, hurried just a little, sliding noiselessly along red carpets between oak-panelled walls, on which hung photos, paintings, and trophies of famous yachts and sailors, but the members, if they were not actually born in those cavernous leather armchairs, certainly exuded a proprietorial air.

*Got it.* I sank into one of the chairs myself, almost enfolded in ancient leather. Usually, I felt uneasy when ensconced in these chairs, comfortable as they were, being served, like some colonial master, by the deferential waiters, but today I was in another world. Page after glorious page of colour photos burst into my consciousness like fireworks. Sunrises at sea, storms, remote tropical islands; Robin Lee Graham and *Dove*, in a thousand settings, sprang back into my life with a detail hitherto unimaginable.

The *Robin* bit took a little getting used to. Instead of adopting it, however, I decided that I would continue calling my hero Lee. That's how he'd introduced himself in Durban, or at least, that's what the media had called him. It was a special link between us, in my mind.

I spent the next few hours transported into the future I knew I belonged in. It was a world structured by *National Geographic's* evocative prose, and illuminated by arresting images. The opening page, for instance, showed *Dove* silhouetted against the sunset, *on a serene and empty sea*, as the caption put it. Superimposed on this glowing sky was the story's title, *A Teenager Sails the World Alone.* The opening paragraph wrote its way straight into my heart. *With genoa and mainsail rigged wing and wing, we sleigh-ride down into the deep trough of a trade-wind sea. Then* Dove *labours up the following crest, and down we plunge again, day after day, my boat and I.*

As shadows lengthened, and the harbour began to smoulder in late evening light, I reluctantly returned the magazine to its rack. The temptation to steal it, stuff it up my shirt, was almost irresistible. Nobody else could possibly appreciate it as much. But morality prevailed, and, bereft, I made my way out into the street. I could not resist one last look at the yacht basin before going home. With a mind filled with images of Robin Lee Graham, palm trees, and foam-flecked seas, the ocean voyaging yachts lying at the pier looked more exciting than ever. I was blind with desire.

Not long after I discovered that *National Geographic* article, the two brothers who'd rebuilt *Sandefjord* and sailed the old ship around the world, Barry and Patrick Cullen, released a two-hour film about their voyage, *Sandefjord, Her Voyage Around the World.* It ran in a local theatre for two years and became another icon of the ocean voyaging life for me. I would have attended every performance if I could, but must have seen it at least 15 times.

It drove my father nuts. To him, a child of the depression, I was just squandering the money I earned from my Saturday job. I felt like pointing out that he squandered his money on cigarettes and alcohol, but valued my ears too much.

The crew introduced the film personally in early screenings. They were large men, with thick, black beards. Afterwards, they stood in the foyer and answered questions, though I was too shy to speak to them. My memory is that they were all seven-foot giants.

*Sandefjord at anchor in Robinson's Cove, Moorea, Society Islands, February 1966. Photo: Sandefjord archives.*

They also produced a magnificent, 12-page booklet, which I still have, its aged pages carefully preserved in a plastic sleeve. I have read it so often I know the words almost by heart.

*It was a desperately sad shadow of the once proud and gallant SANDEFJORD that was found half sunk at her moorings in 1963. Left untended for so long any vessel will deteriorate and SANDEFJORD had almost reached the point of no return when we bought her for the proverbial song. The task of refitting her required almost two years of hard work before she was ready for sea. She was taken from the water, stripped of all doubtful planks and timbers and slowly restored to a state of complete seaworthiness.*

*Finally, in February of 1965, SANDEFJORD was ready. She was provisioned for 400 days and with her complement of five young men and a girl she sailed from Durban on what proved to be the greatest adventure yet. The people of Durban had taken the character and personality of this Grand Old Lady to their hearts and many hundreds lined the breakwater to wish her well. The same people were there 22 months later as she completed her circumnavigation... the vindication of her splendid Scandinavian ancestry.*

*Within hours of sailing from Durban SANDEFJORD and her crew were put to the test as a strong gale blew up from the South, striking the old ship with relentless fury. She rose to meet the challenge, thrilling the crew with her inherent competence. One gale preceded another in this trying passage to the Cape...but once in the Trade Winds of the Atlantic bad weather was soon forgotten and she rushed ahead in idyllic sailing conditions.*

*Through the West Indies, Panama Canal...and on into the mighty Pacific...the greatest expanse of water in the world. SANDEFJORD made her landfalls in the exotic South Seas in much the same way as Cook and other early navigators...there is a legend which says that if you have not arrived in Tahiti by sailing ship, then you have never really been there.*

*Without exception she was well met at all her ports of call. She made friends easily...for herself, her crew...and the country under whose flag she sailed. She was a lucky ship, her people stayed together through thick and thin, as loyal and devoted a crew as any ship could ever wish to have.*

All I had to do now was find my own *Grand Old Lady*, buy it for the *proverbial song*, patch it up and sail away to Tahiti and the islands of Australia's Great Barrier Reefs, just like they did. I returned to the Bayhead and re-examined the old wrecks, but they were, one and all, far less than *desperately sad shadows of former glory*. I would have to find another way.

*Sandefjord cruising through the islands inside Australia's Great Barrier Reefs, August 1966. Photo: Sandefjord archives.*

That summer, I formed a close friendship with one of the voyagers, an American called Keith Kibler, and his companion, Felicia Frey, aboard the yacht, *Korsar*, after striking up a conversation with them on the jetty. I had lost some of my shyness, though it has never entirely disappeared. Felicia was slim and fit, with raven-black hair. She seemed quite a bit younger than Keith. He was the youngest-looking 59-year-old I'd ever seen. Most of the time he just wore a pair of ragged shorts, went barefoot and bare-headed. His hair was greying but his broad chest was muscular and tanned. He always had a generous smile for me.

Like most voyagers, they were interviewed by the local media. The *Daily News* published a photo of them standing on the bow of *Korsar.* Keith was described as a retired stockbroker, fulfilling a boyhood dream, while Felicia was quoted as saying that she'd met Keith in Honolulu several years earlier and decided to see the world with him. I made the mistake of showing the article to my mother, but she was scornful because they were not married. 'You'd better not mention you've been talking to them to your father,' she said ominously.

*Keith Kibler on the bow of Korsar in Honolulu, 1959. Photo: courtesy Keith Kibler.*

*Korsar* was a 39' carvel-planked cutter, built in Germany in 1939 as a racing yacht, sleek and low, with a rakish spoon bow and canoe stern. It had been a very advanced design in 1939, and made the 1700-mile passage from Mauritius in 14 days, the fastest time recorded by any yacht of that size in 1968. Keith told me the boat always took a few days off the other yachts on passages. A distinctive feature of *Korsar* was that Keith used an electrically-driven automatic self-steering device that he'd designed and built himself, called George, with a tiny windvane mounted near the boom gallows. Electric autopilots for yachts, which have now become ubiquitous, were largely unknown before Autohelm introduced its first off-the-shelf model in 1974. Even then, they were not widely adopted by small to medium-sized voyaging yachts for some time, due to concerns about power consumption and reliability.

The first time I talked to Keith, I just perched on the dock, although I did try a little surreptitious peering down the hatch. He eventually invited me on board, but it was too late and I had to run for the last bus that would get me home before my father returned from work. Being late was non-negotiable. But on my second visit, Keith greeted me like an old friend. 'Come aboard,' he said casually.

I clambered down gingerly, trying not to show how uncertain I was, stepping into the cockpit and feeling the boat stir beneath my feet. *I was afloat.* What's more, I got to go inside, after all those months of peering through portholes. *Down below,* as they said in the books. It was my first time on a boat. In the snug cabin, Felicia was putting a few stitches in the mainsail. 'Just overhauling it for the passage to Cape Town,' she said. 'You know what they say, a stitch in time saves nine. Have a seat.'

Sitting there, with a chart of the Cape of Good Hope on the chart table, and the sextant in its box beside it, I felt that my ocean voyaging life had begun. I was flushed with excitement and just a little sweaty from the motion. My tummy gurgled. I hoped I wouldn't be seasick. *In the harbour. The shame of it.*

I remember that chart table. It seemed to me to be the heart of the boat, the repository of all the arcane knowledge one needed to sail safely across oceans. Navigation was by sextant, compass, chronometer, barometer, paper charts, atlases of ocean currents, tide tables, etc. The charts were stowed in a compartment under the hinged chart table top, and the rest of the navigation equipment on surrounding shelves. It was here, I thought breathlessly, that Keith plotted *Korsar's* position from his sextant observations, or deduced the ship's likely position when it was not possible to obtain fixes from celestial bodies, or bearings on land.

Dead (as in ded, diminutive of deduced) reckoning is an art, a fusion of information drawn from various sources like tide tables and current atlases, the ship's course, speed and leeway, plus instinct. The latter is what separates journeymen from brilliant navigators, or did before the age of satellite navigation.

Even good navigators could get into trouble in some areas, like the Tuamotus or Dangerous Archipelago, NE of Tahiti, due to unknown ocean currents and periods of poor visibility. *Korsar* had gone up on the reef there and was lucky to get off. I only discovered this when asking about their large tape recorder, which Keith admitted had not worked since it was inundated with seawater during that incident. He didn't like to talk about it.

Today, some modern boats don't have chart tables. They have navigation stations, with a small area to place laptop computers if you are lucky, surrounded by more instruments, almost, than an airliner's cockpit. I like a proper chart table, with all the information and tools at hand to practice physical navigation.

*Korsar* had been taken as a war prize by the British Royal Navy after the Second World War and used for sail training for some years. Keith bought the boat in England in 1959, and had been voyaging for the last 10 years.

'Ten years,' I exclaimed. 'How many times have you sailed around the world?'

He gave an easy laugh, 'Oh, I haven't. Not yet, anyway. I'm real slow, you see. I like to spend time in places.'

'Too much time,' added Felicia, who was always teasing Keith. 'That's our weakness, you see. We are always missing seasons. I have to nag Keith to pull the anchor up. Likes to drag his heels, this boy does.' Keith just smiled.

*Korsar's* white hull had numerous rust stains on the topsides. One day I arrived at the jetty to find Keith sitting in his dinghy, chopping bits of planking out and gluing in small pieces of wood he called dutchmen. 'Like the patches on a Dutchman's pants,' he grinned.

'Why does it happen,' I asked, "Can't you stop it?'

'It's iron sickness,' he said, making a wry face as he chopped out another piece of wood. 'The boat is iron-fastened, and the rusty fastenings rot the surrounding wood. I'd have to refasten the whole boat to stop it. It just isn't worth it. One day the whole damned thing is going to fall apart.'

I must have looked alarmed, because he chuckled, and said, 'I'm only joking, Graham. There's plenty of life in the old girl yet.'

After that day, I became quite at home on *Korsar*, spending every possible hour I could on board. The summer school holidays happily coincided with the season for sailing around the Cape of Good Hope. As soon as my father left for work, I'd scuttle off to the yacht basin, with the weary collusion of my mum. Felicia showed me how to make popcorn, and how to turn old rope into baggywrinkle. When I first saw that new baggywrinkle, fluffy and bright, lashed to the rigging where it would protect sails from chafing on the wire, my heart swelled. *That very baggywrinkle*, I thought, *which I've made with my own hands, will soon be sailing through the trade winds of the South Atlantic, to the West Indies and beyond*. One day, Keith taught me how to make an eye-splice in a mooring warp, and later I had the thrill of seeing that warp holding *Korsar* to the jetty. It is so easy to be thrilled by simple things in youth. I have to try harder these days.

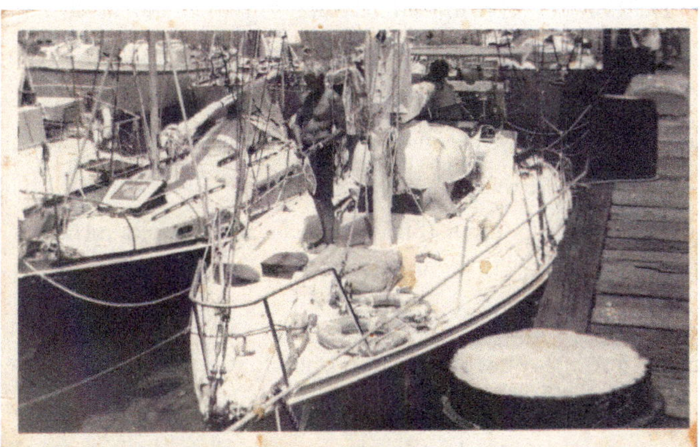

*Korsar alongside the International Jetty, December 1968. Keith is on the foredeck while Felicia leans over the boom. Johann Trauner's fibreglass Folkboat, Lei Lei Lassen, is alongside Korsar, while the Durban-built trimaran, Audacious, is on the outside.*

Like swallows, trade wind voyagers briefly gather in ports such as Durban during the summer months, avoiding cyclones in the tropics, then the season changes and they sail away again. I knew *Korsar* would soon be leaving but didn't want to think about it. The Austrian singlehander, Johann Trauner, who had started his circumnavigation from Toronto, Canada, in 1966, had already left aboard *Lei Lei Lassen*, his fibreglass Folkboat.

*Johann Trauner and El Tigre aboard Lei Lei Lassen.*

Johann was a short, stocky man, with receding ginger hair, not at all like my image of a singlehanded sailor. He had a naughty cat, also ginger-haired, called *El Tigre*, who liked to sharpen his claws on other people's sails. *Lei Lei Lassen* had been moored alongside *Korsar*, and Johann was close friends with Keith and Felicia, but he was diffident with me and I seldom got more than a few words out of him. I identified with him, though, partially because he was a lone sailor on a small boat, and also because he was trying to sail around the world with just CA$1500, to demonstrate that you didn't need to be rich to do it.

This fact had a strong influence on my economic strategy, though I have never come close to emulating it. I like buying books too much, for a start. *Lei Lei Lassen* had the same sort of Perspex viewing bubble on the cabin top that I had seen on *Rehu Moana*. When I mentioned this to Johann, he said it was made for a Spitfire fighter jet. In those days, there was still a lot of surplus World War Two stuff around. The Timm family had also left on their trimaran, *Audacious*, a larger version of Tom Corkill's little *Clipper*, that they had built locally. It was a good name for their boat, I thought, thinking of Tom's capsize.

My heart lurched one day in late January, when I arrived at the dock and saw that *Korsar* had been moved to the outside row of yachts. I knew what that meant.

'Hey!' said Keith, giving me a hug, 'Glad you made it down. We're all ready to go. If the wind swings NE tomorrow, as forecast, we'll be off about 9 am.' He ruffled my hair. 'Don't look so sad, my boy. Next time we meet, you'll be on your own yacht and we'll have a real party.'

I grinned. 'Sure thing, man,' I drawled. I'd been cultivating an American accent all summer.

At 0900 the next day, I was standing on *Korsar's* foredeck, feeling it lift to occasional surges from ship wakes in the outer harbour. I tried to imagine what it might be like, clambering around that deck on the ocean. Keith had started the engine and I could feel it rumbling beneath my feet. Then the boat was moving. At the last possible moment, I jumped off, scrambling across the decks of other yachts rafted to the jetty, heedless of cries to watch out for fresh paint. My plan was to run to the end of the concrete yacht mole, which jutted well out into the harbour, where the water sucked and swirled around you and it felt like you were almost afloat. I often went out there, though it was a bit isolated, and like all such places in South Africa, carried a frisson of danger. There wasn't a minute to spare. I ran like a dervish through the car park and out along the mole.

*Korsar* was sliding past now, agonisingly beautiful, low, white and sleek, its elegant spars stencilled against the summer-blue sky, sweet canoe stern balancing the lean, hungry bows. The familiar figure of Keith was at the helm, and Felicia stood on the foredeck, coiling mooring warps. She raised her hand in farewell, and I thrashed my arms in a whirlwind, then cupped my hands to my mouth. 'See you down the line,' I yelled, as I'd heard other voyagers say.

'You betcha,' Keith yelled back.

A few weeks later, I received a letter from Felicia. *Korsar* was in Cape Town, after making stops at East London, Port Elizabeth and lastly Knysna, where the difficult, shallow, rock-strewn entrance had frightened the daylights out of them. They'd also weathered a gale the night after leaving Durban that tore the genoa. I had been thinking of them that night, as I lay in my safe bed listening to the wind moan through the trees outside my window. Johann and the Timms were also in Cape Town, she said. Felicia's letter gave me a strange feeling, as if I was peering into a room that I longed to enter.

Several months later I heard from them again, in a letter postmarked Rio de Janeiro. I remember the thrill of receiving that letter, with its exotic foreign stamps. They had an easy passage across the South Atlantic via Saint Helena Island, with just one day of bad weather soon after leaving Cape Town, and a couple of wet days approaching Rio. They waited several days at Saint Helena Island for Johann, as arranged, but he had not arrived, and they were concerned about him. He must have missed the island, not unknown in those days of sextant navigation, or decided not to stop, because official records show that he completed his circumnavigation later that year.

After reading about *Jester*, meeting Johann, and reading *Sailing into Solitude*, the book Val Howells wrote about his transatlantic voyage aboard the Folkboat, *Eira*, in the 1960 OSTAR, a plan formed in my mind. I would go to England after leaving school, buy a Folkboat, which in those days were selling for £1000, convert it to junk rig, compete in the 1972 OSTAR to establish my credentials, then continue on around the word six or seven times.

*Tim McCloy and his Jester sistership, China Blue, off the west coast of Scotland. Photo: courtesy Tim McCloy.*

People in the yacht club might laugh at *Jester*, and Blondie Hasler himself said the name was inspired by the boat being a bit of a joke, but it was a brilliant concept really. There are a couple of sisterships that I know about, *Misty* and *China Blue*, both of which have completed long ocean voyages, with the latter being an exact replica of *Jester*. *China Blue* was sailed out to Kenya from England by a young couple, before being bought by Kenyan sailor, Tim McCloy, who brought it back to England and participated in a number of *Jester Challenge* events, plus making a circuit of the North Atlantic. Tim lives aboard and was last heard of in Portugal.

I'd been told by the yacht club pundits that the Vertue was too small for ocean cruising, so I didn't ask what they thought of the Folkboat. If they thought the Vertue was too small to swing a cat in, I assumed they'd think you couldn't even swing a mouse in a Folkboat. I didn't know then that the term referred to the whip known as a cat o' nine tails, but they seemed to me to be the sort of people who might just be mean enough to swing some poor cat around by its tail.

# Chapter Four

# Maiden Sail

The first half of 1969 was an exciting time in the history of singlehanded sailing. Despite a number of problems, Robin Knox-Johnston had arrived back in Falmouth on 22 April aboard his 32' carvel-planked ketch, *Suhaili,* having sailed non-stop and solo around the world, winning the Golden Globe Race. In the same event, Bernard Moitessier astounded everybody by going around one-and-a-half times on his bullet-proof steel ketch, *Joshua*. As Moitessier put it, if he didn't break his neck in a pitchpole, *Joshua* would always make port somewhere, even without masts. It seemed at the time that nobody would ever surpass his 37,455nm nonstop passage, but of course, people love a challenge.

Both men were well known in Durban. Moitessier spent a year there in *Marie Theresa II* in 1955-6, and Knox-Johnston lay over on *Suhaili* for several months in 1966. There were stories about them in local newspapers, and conversations abounded in the yacht club, where their friends remembered them well. I lapped it all up, but could not wait for the following summer, when I would have finally finished school, and the new crop of voyaging sailors would arrive. I was planning my escape.

Soon it was mid-winter. As always, there was little happening in the Durban yacht basin. Fred and Virginia Davenport were still there, with their 13-year-old daughter, Circe, on *Karen Margrethe*. This American-flagged, 46' Danish ketch, carvel-planked and canoe-sterned, had once been a fine vessel, but was looking neglected, and the masts, which had been removed, were lying on the ground up near the yacht club. Fred and Virginia had been there so long, even before *Dove* sailed in, that they were almost locals.

I went down once a week or so and talked to Fred, who was always affable, in that American way that delighted me, but Virginia was getting more frazzled every time I saw her, perhaps due to her inability to motivate Fred to refit the boat. She seemed to have started hoarding newspapers. Meanwhile, Circe and I eyed each other warily, two teenagers wrestling with their hormones, strangers in their own bodies. Other than that, it was mostly a time for reading and dreaming.

Then, one night during the July school holidays, the phone rang. 'Have you seen the new yacht on the dock at the T-Jetty?' a school-friend, Hilton Petters, asked. I hadn't. The T-Jetty was a commercial dock some distance from the yacht basin, where *Sandefjord* had been lifted out for its refit. As early as possible the next morning, I rushed down. There was nothing there.

I cast a wild eye across the harbour. Nothing. There were a few places where it could have gone, but the most likely was the yacht mole, or basin, as you might more correctly describe it. It was also the closest. *Idiot.* I'd run right past there, but for the first time in my life, at least since I had become obsessed with ocean voyaging, I hadn't looked in. My feet pounded back down the Esplanade. *Yes.* There were only a handful of boats on the International Jetty, and I could see, from several hundred metres away, that there was a new mast among them. I knew every mast in that harbour.

The new yacht was called *Clipper*, which surprised me. I had no idea how common that name was. For a moment, I thought it might be Tom Corkill, until I saw a muscular, balding man tuning the rigging, who turned out to be a South African called Clive Rouse. *Clipper* was a 28' sloop, white with gleaming varnish, with just the right amount of freeboard, low and sleek, a little bit like *Korsar*.

And it had a self-steering gear just like *Dove's*. Except it wasn't exactly the same, I discovered later. *Dove's* windvane drove an auxiliary rudder on the transom, with the main tiller lashed, while *Clipper* had a windvane that turned a trim tab on the trailing edge of the rudder, and the water flow over the deflected trim tab turned the rudder. At least *Clipper's* windvane was similar in shape to *Dove's*, but even that changed some months later, when Clive adopted Blondie Hasler's design of an offset windvane connected to the trim tab by a differential linkage, to eliminate over-steering, and also used Hasler's distinctive windvane design.

By this time, wind-driven self-steering gears had become common, were almost a functional signature of long-distance cruising yachts. My generation grew up with them, the way kids today have only ever known the continuous connectivity of smart phones, but before 1960, wind-driven self-steering had been virtually unknown. Apart from lightweight marine engines and synthetic sails, they were one of the most significant developments of the Slocum era.

There were a few old-school voyagers in the late 1960s who resisted the proliferation of self-steering gears. They claimed that these devices had caused an influx of first-time sailors who were crowding the best harbours and anchorages. The inference was that this new generation was too soft, lacked proper seamanship, and that there were too many of them. Today, even windvanes are seen by many voyagers as old-school. Increasingly, modern yachts rely on autopilots, especially since so many of them are unsuitable for mechanical self-steering systems.

Clive built *Clipper* in his mother's backyard in Johannesburg, 400 miles from the sea. Working part-time, in the evenings and weekends, while holding down a full-time day job, he finished *Clipper* in two years, a phenomenal feat.

*Clipper under construction in Johannesburg in 1968. Note Clipper's rudder hanging up behind Clive. Photo: Clive Rouse.*

He'd trained as a die maker, and the quality of construction was magnificent. The hull was double-planked fore-and-aft, glued and riveted, and he had also doubled the number of floors, fitting one alongside every steamed rib, thus making the boat phenomenally strong.

'Hello, Sonny-Jim,' he said, looking up with a big smile, as I stood gazing at this new boat to drool over. Launching a boat is always a happy occasion, but I was soon to realise that Clive was always smiling; smiling and whistling. He was a quiet, steady, self-confident man. 'Come aboard if you like,' he added. 'I'm just tuning up the rigging and then I'm going to take her out for her maiden sail. You can come with me if you want.'

'Oh, wow! Thanks. Can I help?'

'Sure. You can undo those shackles on the cabin top and pass them to me as I need them.' As I undid the first one, *Clipper* gave a tiny lurch, most likely from the wake of a distant tugboat in the harbour, and I dropped the pin. It bounced on the deck and went over the side with a decisive plop. I blushed, but Clive only laughed, saying, 'Never mind, I've got several spares.'

*Clipper was built phenomenally strong. Note the closely spaced ribs, with a floor adjacent to every one, twice the number specified by the designer. Photo: Clive Rouse.*

I was acutely aware, as I bumbled and fumbled, that I was more of a hindrance than a help, but Clive never stopped smiling and whistling. We did not quite finish the task, so didn't go for a sail, but he promised to take me the following morning.

The next morning, I scampered off from home as soon as I could. There were certain protocols in our house. If I left before my father, I'd have to ask him for permission, and I was worried he'd say no. His fear and hatred of the sea made him uneasy with my interest in sailing. I sometimes felt like asking if he'd prefer me to take up ballet after all. I was terrified that Clive would set sail before I arrived.

*Clipper alongside the International Jetty, second row of yachts, in the middle. Photo: Clive Rouse.*

'Ah, there you are,' said Clive. 'I was beginning to think you weren't coming.' *Clipper's* engine was running and he'd already taken off most of the dock-lines. We cast off the remaining lines and *Clipper* slowly backed away from the jetty. The bows turned towards open water and off we went, chugging down the channel into the bay. I felt a lightness of spirit, enriched with anticipation. After more than 55 years, I still experience this uplifting feeling whenever I cast off the dock lines.

*Clipper* was emerging from the channel into the open bay, which shimmered like gunmetal in the early morning light. In the distance, I could see smoke streaming lazily from the red and black funnel of a ship moored at the T-Jetty. It was going to be a beautiful day.

I felt a twinge of anxiety. It was the first time I'd been outside the yacht basin in anything smaller than an ocean liner, and I hardly remembered that, since I'd been three years old. It looked like a long way to swim. I knew that I was prone to anxiety, but I'd believed that things would change when I had a good yacht under me. I trusted Clive, and *Clipper* was the best boat ever built, and I knew the West Indies were a long, long way across the sea, but experiencing it all was something else. Just then, Durban Bay looked as big as the Atlantic Ocean.

Once clear of the channel, we set the sails. There was another friend of Clive's on board, and I gratefully perched on the rail, watching him run around the deck pulling ropes. In my excitement, I couldn't remember what a halyard was, or a traveller. I remembered that sheets were the ropes you used to pull sails in with, but didn't know which ones they were among the jumble in the cockpit. Sails were flapping, ropes were rattling; confusion reigned, in my head anyway. Clive was whistling happily, with a smile that was even bigger than usual, if that was possible, as he stood proudly at the tiller.

*Clive whistling happily as he helms Clipper off Durban.*

His friend seemed perfectly relaxed. Then Clive switched the engine off, *Clipper* tipped over on one side, and everything went quiet, except for a hiss of water out the back. We were sailing!

Clive looked at me and said, 'Here, Sonny-Jim, take the tiller, I need to pee.'

'I...I...,' I stammered.

'Don't worry,' he said, giving me a pat on the head. 'You'll be OK. Just point *Clipper* towards that ship over there. If the sails flap, ease off a bit to port; that's the left side. The other side is called starboard. Whatever you do, don't tack suddenly, that is, turn up into the wind so that *Clipper* tips over the other way, or I might pee in your ear.'

I clutched the tiller between my clammy hands, trying not to panic. The bay suddenly seemed very small, and *Clipper* was hurtling to destruction against the rust-streaked sides of a large, black ship. Distracted by my fears, I did not notice that the boat was slowly edging up into the wind, sails shivering as their angle to the wind became too slight.

'Bear away,' called Clive, motioning the direction in which to go with one hand, the other still engaged in men's business, feet balanced precariously on the toe-rail. I shoved the tiller over.

'Wrong way,' he yelled, making a wild grab for the backstay to save himself, as *Clipper* swung up into the breeze, sails flapping. I shoved the tiller the other way, *Clipper* heeled to the breeze again, the sails fell quiet, and the gurgle of water running past the hull resumed.

'Good,' said Clive, 'I didn't even pee down my leg. Now remember, when you push the tiller one way, the ship goes the other. It's the opposite to what happens when you turn the wheel of a car.'

I was thrilled with this perversity. I still had the tiller, however, and that ship was looming. 'What about that ship?' I asked nervously. Just then, *Clipper* seemed so huge, so powerful, seemed to be moving so fast. Surely it needed more space to turn.

'Don't worry,' Clive smiled, 'that ship is still miles away, but it's a good idea to be cautious.' He patted me on the head and I would have purred if I'd been a cat. He continued, with a slight smile on his face, 'You never know what might happen next on a boat. The wind might suddenly change or stop. You've always got to think one step ahead. You need a flexible attitude to be a good sailor. Not too many bank managers sailing around the world.'

At this last comment, his friend roared with laughter. There was obviously some private joke being told here, but I grinned too. I was never going to be a bank manager. Clive took the helm to tack the boat around. I hauled the jib in on the other side, having worked out where the sheets were, and off we went, back out into the bay. He gave me the tiller back and I proudly steered the boat on a straight course. I was getting the hang of this.

Then a gust came, and the boat heeled over until I was sure *Clipper's* mast would smack the water. My feet were slipping on the cockpit floor, and I struggled to hang on to the tiller. *Clipper* was only four feet longer than *Dove*, but I was overwhelmed by its energy.

A doubt flicked through my mind. *Will I ever be able to sail singlehanded?* Then Clive laughed, a clear, gleeful laugh that blew away my fear. 'Look at her go,' he shouted. It was the most glorious day I had ever known.

Clive set sail for the Seychelles soon after, but a few days later I was astonished, and delighted, to discover *Clipper* back at the jetty. *Clipper's* crewmember had lost his nerve, the stove and the self-steering gear had malfunctioned, and Clive had decided to stay an extra year, go back to work and sort things out.

He invited me to sail with him on weekends. When I mentioned I was planning to go to New Zealand after finishing school, he said, 'Hey, let's sail there together. I want to go up to the Seychelles next May, then we can go on down to New Zealand.' I nearly swooned with pleasure.

Being the sort of person who can turn his hand to anything and do it well, Clive was a natural seaman, but he also had the benefit of sailing with Bob and Nancy Griffith on their 53' ferro-cement cutter, *Awahnee II*, from Durban to Valparaiso, Chile, via Patagonia, in 1966. Bob was one of the legends of the Slocum era, a tough, strong-willed, resourceful and fiercely independent sailor, who found his perfect match in Nancy, a competent, adventurous sailor in her own right.

*Awahnee II sailing off Honolulu. Photo: Bob Griffiths archives.*

Between them, Bob and Nancy honed the skills of making ambitious voyages with limited resources, often with green crews. Many young sailors learned the foundations of good seamanship under Bob's command, if they could take the discipline.

Bob, with much input from Nancy, wrote a book that quickly became a classic of the genre, *Blue Water: A Guide to Self-Reliant Sailboat Cruising*, offering advice in everything from boat design, sail handing, anchoring, seamanship and provisioning. In recent years their children put together a magnificent documentary film, *Following Seas*, from footage taken by Nancy Griffith during their travels. It is available on *Vimeo*.

When Clive returned from Valparaiso, he knew exactly what boat he wanted. He'd found the design, a Clyde Six-Tonner, from the board of English designer, Alan Buchanan, in a magazine of Bob's, and Bob had given it the seal of approval.

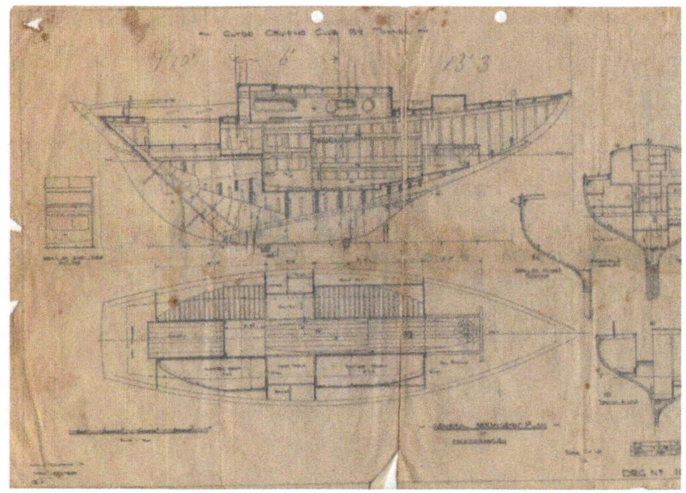

*Clipper's profile construction drawing. Note the perfect accomodation layout. Photo: Clive Rouse.*

Like *Awahnee*, the design had a cutaway bow profile, something Bob believed was crucial in a sea-kindly hull. *Clipper* resembled *Awahnee* in another way, in that it was a very simple, rugged boat. *Clipper* had hanked-on headsails, a hand-cranked Stuart Turner petrol engine, and oil lamps. There was no battery or electrical system.

For the rest of the year, I sailed on board *Clipper* most weekends. Initially, when it was just the two of us, sailing around Durban Harbour during my school holidays, all went well. Clive taught me how to hoist the sails, and to sail the boat on and off a buoy in the harbour, which we'd tie up to for lunch. We feasted on stale bread and cans of corn, washed down by mugs of sweet tea. Never had a repast tasted so delicious.

Clive became the most important person in my life, and I hung off every word he uttered. I considered *Clipper* to be the epitome of everything desirable in yacht design and construction. I still have a deep emotional bond to *Clipper*, and to this type and size of yacht, with its long keel, cutaway profile, transom-hung rudder and utter simplicity.

*Clipper returning to the jetty after its safety inspection. I suspect that Clive knew more about yachts and the sea than the inspector.*

But then I went back to school and Clive got his harbour clearance to sail *Clipper* out on the ocean. You couldn't sail out of the harbour whenever you liked in Durban. The boat had to be inspected and the skipper pass a test. Even then, the skipper had to call the harbourmaster for permission each time the yacht went out. There was a rumour that the harbourmaster had a notice on his wall that said no yacht could go out to sea if the wind was over 35 knots, unless it was a Vertue. I doubt the truth of that, even if he loved Vertues, but it might have been his whimsy.

Once we started sailing in the ocean, always with other crew aboard, I discovered that I was prone to seasickness. I'd lie on the side deck with water sloshing over me, being trampled by the feet of the other crew, oblivious in my misery, but as soon as we got back into harbour I'd recover instantly and wolf down a cold pork pie, or whatever I had brought for lunch. It made Clive's crew laugh, but I was a hindrance out at sea, and would have been a liability in bad weather. 'You'll need to sort that out,' Clive said, 'if we are going to sail to New Zealand.' He was immune to seasickness.

There were some good days, when the sea was calm and I wasn't sick, but I was also clumsy, always dropping or breaking things. Once, Clive came up on deck with several shackles that were minus their pins and threw them over the side. He said he'd been saving them in case I dropped a few shackles instead of the pins, but I never did. He shook his head, to the good-natured laughter of the crew.

*There were some wonderful days when the sea was calm and I wasn't seasick, allowing me to hope that perhaps I might get to sail away with Clive after all.*

There was another reason why sailing with Clive seemed unlikely to eventuate. White South African males under 35 needed to complete compulsory military service, or obtain an exemption from the government, for a passport to be issued, and the exemption did not look like it was going to happen. I'd written to the authorities, saying I wished to sail away with Clive, and been rebuffed. I then wrote an angry letter denouncing Apartheid, and saying I would not serve in the South African military. To be precise, I said that if trouble broke out, they'd be the first ones I'd shoot. They responded by arresting me, beating me up, then telling me that if I did not present myself for call-up on the appointed day in July 1970, they'd jail me and throw away the key. Apparently, there were two Jehovah's Witness boys, declared conscientious objectors, who'd been in jail for seven years already, refusing to concede.

If there was any chance left of sailing to New Zealand aboard *Clipper*, it probably ended one Sunday, soon after the new crop of voyaging yachts began to arrive. Clive wasn't sailing that day, having just repainted *Clipper's* decks, and they gleamed spotlessly white in the sunshine. I started talking to Fred on *Karen Margrethe*. He was an enigmatic character, who spent his days wandering around with a camera, photographing anything that moved. He had no visible means of support, as far as I could see.

The Immigration Department, worried that he might be an American spy, perhaps, had just given him an ultimatum to leave in three months. Usually, they hung onto every white person they could get. So here was Fred being kicked out, and me trying to get out. 'Let's swap,' I said, and we both laughed.

Virginia, Fred's wife, had enlisted my help in trying to motivate him. 'Ask him when he's going to put the new masts up,' she suggested. The old masts had rotted on the ground years ago. The new ones, rigged and ready to hoist, had been lying on the jetty for weeks, getting in everybody's way. Fred and I were leaning against a railing on the jetty, and Circe sat on a wharf pile nearby. She was just 14 but had the body of an 18-year-old. She was deeply embarrassed by her sexuality, and I was trying to run away from mine, so I was always flirting with her. She hated boats, hated sailors, hated boys, as far as I could tell. I could have been wrong.

'Say, Fred,' I asked, as casually as I could, 'what's holding these masts up?'

He just grinned. 'Nothing's holding them up, kiddo. They're lying on the ground.' He twirled the lens of his camera, pointed it down at my sun-tanned legs, which stuck out of the brief white shorts I usually wore, and took a photo.

'What do you want a photo of my legs for?' I protested, amused.

'Circe might like one for her album,' he replied. He was always teasing her about being attracted to me.

She blushed, jumped up and stomped off, but soon came sidling back.

'So, what are you planning to do next?' I asked, trying to advance Virginia's agenda.

'Those bureaucratic bastards have thought up another way of making my life difficult,' he replied. 'They've instructed me to remove my boat from the yacht basin and take it up to the Bayhead, but that's the last thing I'll do.'

I might have loved the Bayhead, with its old wrecks and straw dreams, but Fred and Virginia were not so impressed. They had no transport, and it was miles to anywhere. Just to buy a loaf of bread would have been an all-day excursion.

'What are you going to do, then?' I asked.

'I'm not going,' he replied, to my delight. He had thumbed his nose at the authorities for years, probably the real reason he was getting booted out. 'You can give us a hand to load these masts aboard,' he continued, 'and we'll go and anchor just outside the yacht basin, beyond the mole. It's pretty exposed to the south, but we'll get some protection from the sandbanks at low tide, and we'll still have access to the club and town.'

We loaded the masts aboard with the help of another young South African, Pete, and then motored slowly out to the chosen spot. Pete was a couple of years older than me, and had a little boat with a snug cabin that he'd built with his father. He appeared to lead a happy, independent life, and was always friendly. I envied him his freedom, and the way he dressed, in torn blue jeans and unbuttoned shirts. My parents would never let *me* dress like that.

Just after we'd finished laying the second anchor, a ferocious southerly gale began. In no time at all, the old ketch was pitching into steep little waves, straining to its cables. But the anchors appeared to be holding, so we retreated below into the cluttered cabin, clothes soaked and teeth chattering. Fred gave Pete and I some of his dry clothes to wear. 'Can't get you ashore in this,' he commented. 'Guess you'll just have to stay for supper. Maybe even the night.'

I felt a rising panic. I had never yet spent a night away from home. Clive had suggested staying on board *Clipper* a couple of times, but I always made excuses, even though I longed to, because my parents wouldn't approve. I'd looked forward so much to spending my first night afloat, but not like this. My parents would think I'd disappeared, run away or been murdered. People were always being murdered in South Africa.

There was nothing I could do, and nothing I felt I could say, so I just snuggled down. The kerosene lamps were lit, Virginia put on a big pot of stew, and we all squeezed in around the table. It was thrilling, in a way, but my anxiety was like that famous worm in the rose. The gale continued to blow. I was expecting a harbour police launch to crash alongside at any moment, but none came, and neither did the gale abate. Pete and I curled up alongside each other on the cabin sole and slept through the night.

I awoke in the grey light of dawn and realised the storm was over. I lay there for a while, listening to Pete's deep, even breathing, and to the delightful sound of water chuckling against the hull. I felt an urgent need to call home, however, so coerced a reluctant Pete to wake up and row me ashore to the end of the concrete mole protecting the yacht basin. Then I ran along the mole in the early morning chill towards the yacht club, from where I phoned home.

My father answered, sounding weary and strained. Usually, he never answered the phone, even though most calls were for him. Later, my mother said they were convinced I was dead, and when the phone rang, they thought it was the police advising that they had found my body. My parents had scoured the streets for much of the night. Eventually they remembered the name of Clive's boat, and located *Clipper* tied up on the inside row at the International Jetty. It was low tide, and *Clipper's* freshly-painted decks were far below the dock.

After calling out futilely for a while, their voices lost to the gale, they found a large, heavy, indescribably oily rope, and dropped it onto *Clipper's* deck. Clive heard that all right, but he was unable to help them. Nor was he able to get the oil stains out of his new paint, or quite forgive me. I wasn't too popular at home either.

# Chapter Five

# The Debuntante's Ball

Once I finished school, I spent my days hanging out at the jetty, where new ocean voyagers arrived almost daily. I should have found a job, since I had no intention of going to university, to my parents' disappointment. I had been their brightest boy, and they had high hopes for me, but because I didn't bother to study after discovering Robin Lee Graham and ocean sailing, I'd failed my final exams, apart from English and History, in which I just scraped a pass. Looking back, I am surprised my parents did not confront me, either about my academic failure or my lack of work ethic, but I think they realised I was troubled and gave me some space. These days, school councillors would be asking questions as well.

I was often on the jetty when yachts came in from Mauritius, Madagascar, or wherever. There is an aura surrounding a yacht at the end of a long ocean passage. The decks are clean and uncluttered, scoured by wind and sea, and the crew look much the same, with an open, slightly dazed look on their faces. It soon disappears in the grime and grind of shore life, but for a brief moment it is like looking into their souls.

I befriended as many of these sailors as possible, listening enthusiastically to their stories and gaining an insight into the realities of ocean voyaging. Some of their sagas were more complex, and occasionally darker, than those I had been reading about, but they had all crossed at least one ocean, faced at least one gale, and the solitude of the ocean. I was in awe of them all.

Croix Grut was a quietly-spoken, gentle soul, a short, slightly-built 51-year-old dairy farmer from New Zealand, sailing around the world with his wife, Merle. Each of his cows had a name, he said, to which it responded, and a unique personality. He was relaxed and unpretentious, with a scrubbed appearance, as was his yacht, *Iorana*, a recently-launched 36' ketch, designed by a well-known New Zealander called Woollacott. This strip-planked boat was a work of art, professionally built with full-length kauri planks. It had a white hull and cool green decks. 'The final price still makes me wince,' Croix laughed, but his pride in *Iorana* was palpable.

*Iorana on Wilson's Slipway, Durban, December 1969.*

Once, when I was on my way down to the harbour, I spied him taking photos of an Indian flower seller in the market in Gray Street. The market was in an Indian business precinct, for Apartheid was also a commercial policy, and I always felt a sense of trespass there. I sneaked up behind Croix and clapped a hand on his shoulder. In my best Afrikaans accent, I said, 'Excuse me, Meneer, but you cannot take photographs of these flower sellers. It is illegal.'

Croix leapt like he'd been stung by a bee. 'Oh, is it?' he squeaked in a dismayed voice, turning around red-faced, to see me, helpless with laughter, staggering around on the pavement behind him. He stuck out his tongue.

He was the first sailor I met who admitted to being enchanted with the classic, ocean voyaging stories I loved, admitted that he was sailing in the footsteps of his heroes. Most voyagers I had spoken to, with an ocean or two behind them, positioned themselves as fellow travellers rather than pilgrims. All that adulation, they seemed to imply, was for piddlers in short pants, like me, though some of them eagerly borrowed my latest titles.

Croix's shelves were crammed with the same books I was reading, and we had many happy discussions about our favourite authors. I loaned him my newly-acquired copy of Bernard Moitessier's *Cape Horn: the Logical Route*, and we agreed that Moitessier's lyrical writing perfectly captured the poetic essence of ocean voyaging. The following excerpt was my favourite passage in the book:

*What did we do all the time on this uneventful passage? Nothing, absolutely nothing, except live in communion with our boat, which was sailing happily toward the Marquesas, the Tuamotus, Tahiti, with plenty of wind, wind, wind. One must have sailed far away from any land to know how full and rewarding a day at sea can be without anything happening. It is so different from coastal sailing where one is always a little apprehensive, tense, prepared for anything. Out there, in the open sea, we were prepared simply to live.*

*Francoise spent hours on deck, looking at the white foam running past the boat and giving names to clouds. She saw only three rainbows at the bow because the sky, though very clear, was nearly always dotted with large cumulus cloud. I often sat beside her, watching the foam run past in an endless string of bubbles.*

*But my favourite place was on the pulpit where I spent long moments in a state of semi-hypnosis. It was here that I could really feel the sheer power of Joshua, this harmony of strength and gentleness which emanated from the bow and enveloped the whole boat like a halo. A halo of seven colours...including the red...*

*It was here that I could best hear, with all the fibres of my being, the song which the bow had been singing ever since Martinique...give me wind...I shall give you miles...thousands of miles...thousands of miles...*

*And felt utterly at peace with myself, with Francoise, with my boat...*

Furthermore, Croix knew John Guzzwell. John was married, Croix told me, with twin sons, and sailed a 45' cutter called *Treasure* that he'd built in England. He was now based in the Bay of Islands, New Zealand, a place he'd fallen in love with during his circumnavigation on *Trekka*. This information made me feel like I was at the gates of the pantheon. *John Guzzwell sat here, on this very seat in Iorana's saloon!* I reverently stroked the upholstery.

There was a rock-fast steadiness in Croix that soothed the tumult in my mind, and our friendship quickly deepened. We remained very close until his final days, several decades later. Croix's approach to ocean cruising was quiet and methodical, and *Iorana's* voyage around the world was stately, predictable, and trouble-free. If he had any demons in his soul, he seemed to have tamed them long ago.

*Taking Iorana around to Wilson's Slipway. Croix is on the left, while Merle stands to the right. Note the absence of a spray dodger over the main hatch, or other gear.*

Harry Gilbert, the owner, designer, and builder of *Kelasa,* a 36' double-ended, carvel-planked, gaff cutter from Victoria BC in Canada, certainly had a few demons lurking in *his* bosun's locker. He arrived in Durban on Christmas Day, 1969, along with his Australian mate, Adrienne Matzenik, who was about the same age. Aged 45, Harry was one of the most unforgettable voyagers I ever met, a half-crazy, tragic hero, who desperately pursued his beautiful story to the end.

They arrived in port on Christmas Day, 1969, after a marathon passage from the Seychelles. *Kelasa* was a slow boat at the best of times, but on this passage the flax mainsail was rotten, and Adrienne had spent much of the time coming down the Mozambique Channel lashed to the boom, resewing it. To add to their woes, the temperamental old petrol engine proved hard to start as usual, and they were nearly swept past Durban by the current. Several days after their arrival, they still looked stunned.

*Kelasa* had massive scantlings. The boat displaced 22 tons, almost one and a half times the displacement of *Spray* for the same length on deck. It had 19" of freeboard amidships and 8' draft. Harry had not intended the boat to float so deep, and the lower sponson was just inches above the waterline. He suffered from anxiety, and couldn't help beefing everything up during construction. The ballast keel was concrete and there was more in the bilges. I have no idea what *Kelasa's* ballast ratio was, but suspect it was low.

*Harry Gilbert and Adrienne Matzenik on the deck of Kelasa shortly after arriving in Durban on Christmas Day, 1969. The well-known Point Yacht Club is in the background. Photo: courtesy David Matzenik.*

Reminding me of *Moonraker*, with its working boat style, long bowsprit and low freeboard, *Kelasa* evoked a bygone era. Below decks, the varnished, keel-stepped mast dominated a saloon framed by equally massive deck beams, hanging knees, and panelled bulkheads. The sextant box was secured on a shelf alongside the mast. Joshua Slocum was Harry's hero, and Josh would have been right at home aboard *Kelasa*.

Although Harry had no previous boat-building experience, he designed and built *Kelasa* from scratch over a period of seven years, driving taxis at night, working on the boat during the day, and catching no more than a few hours of sleep here and there. Perhaps that was why he now liked to get eleven hours sleep a night, come hell or high-water, according to Adrienne.

*Drying the rotten flax mainsail a few days after arrival.*
*Photo: courtesy David Matzenik.*

*Kelasa alongside the International Jetty, Note the massive rudder cheeks, indicative of the rest of the boat's scantlings.*
*Photo: courtesy David Matzenik.*

Harry said the name *Kelasa* was Greek for 'little ship', but some of the other voyagers jokingly nicknamed the boat, *Colossal*. Not to Harry's face, mind you. He was a fearsomely proud man with a flashpoint temper, and never more so than when *Kelasa's* honour was concerned. He had the build of a lumberjack, with square jaw and angular features, and looked like he could dismember a grizzly bear. His temper, and his ability to talk endlessly, in great detail, about all manner of obscure mechanical topics, had gained him something of an eccentric reputation.

*Harry loading stores aboard in Rabaul, PNG. He was a strongly-built, imposing figure of a man. Photo: courtesy David Matzenik.*

He broke another taboo as well, enthusiastically describing the storms *Kelasa* had survived. The other voyagers played down heavy weather experiences, and were dismissive of heroics. Skippers, in particular, were expected to be laconic, it was part of the culture. Because of this, they nicknamed him *Hurricane Harry*. But he had crossed two oceans, one of them singlehanded, and when Harry talked about seamanship they listened. I still have a treatise on navigation he wrote for me.

Adrienne was also a character, and a striking, auburn-haired beauty. When she worked on deck in her black string bikini, not much work got done on nearby boats, or the women would find errands for their partners to run. A passionate socialist in her youth, who'd grown up in the rougher suburbs of Sydney, she had an irreverent spirit that astonished and delighted me.

*A jubilant Adrienne points astern at Sydney Heads as Kelasa sets sail for New Zealand in the summer on 1966-7. Note the staysail sheet to the tiller via the port quarter. This was Kelasa's only means of self-steering.*

The affection she felt for Harry when they left Sydney had faded by the time *Kelasa* reached Durban. On the first occasion I spoke to her, I made the mistake of assuming that Harry was her husband. 'He's not my husband,' she retorted vigorously. 'If he was, I'd have brained him long ago.' She openly expressed these sentiments in front of Harry, as if to goad him.

'Yes, Dolly,' was all he'd say, in a resigned tone. He seemed somewhat in awe of her.

When alone with me later, he said, 'She's not an easy woman to live with either, you know. When she's got it in for you, she's as fierce as a Bengal tiger. Don't ever cross Adrienne or you'll live to regret it.'

His stories about Adrienne were a mixture of admiration and exasperation. 'She's as tough as nails. Once we had a rat on board, and she grabbed it with her bare hands, rushed up on deck and threw it into the sea.' He also mentioned how she'd seen *Kelasa* out of the window of her Elizabeth Bay flat in Sydney, come down to take a closer look at the boat, and decided then and there that she was going to sail away with him. Harry, who came from a deeply religious family, and was painfully shy, had never even had a girlfriend, despite having just turned 40.

*Adrienne sewing the rotten, flax mainsail in the Mozambique Channel, en route from the Seychelles to Durban. Photo: courtesy David Matzenik.*

He showed me a photo he'd taken of Adrienne when she was lashed to the boom, sewing the rotten flax mainsail in the Mozambique Channel. 'She was swearing at me like a trooper,' he said, 'saying that if I had time to take bloody pictures, I could come up there and help sew the sail.' He had chosen to retreat below and continue his struggle with the recalcitrant engine.

Some of Adrienne's stories were unrepeatable, at least publicly, and others were hair-raising. I wasn't sure whether to believe all of them, since it was obvious that she loved nothing better than telling an entertaining yarn, but I was to remember one, some years later, in less amusing circumstances.

'Harry has never hurt me,' she said, 'but he has violent fantasies at times. Once, after we'd had a bit of an argument at sea, he called me to come up on deck. As I came up the companionway, I saw him standing on the bridge-deck, holding a hammer. It looked like he was going to hit me with it. *Don't you dare,* I shouted. I'd worked out by then that the way to handle him is to be extremely firm, like dealing with a five-year-old child. He immediately put the hammer down and stuttered a denial. He always stutters when he feels guilty, or is trying to hide something.'

'Why do you stay with him, then?' I asked.

'Well, I feel loyal to *Kelasa*. I fell in love with the ship the moment I saw her. She's the real hero of this story. She's saved our lives again and again in the most appalling weather. Harry treats her badly, but she's never let us down. I promised Harry that I would help him get her back to Canada.'

After *Kelasa* had been in Durban for almost a year, the Port Captain requested Harry to move to Dead End Creek, in the Bayhead area, to make room for more recent arrivals. It was out past the dry docks where I had gone to visit *Dove*, and I took to riding my bicycle there once a week. I loved sitting in the magnificent ambience of *Kelasa's* cabin and listening to Harry's stories. He might have been a bit crazy, but he was the archetypal Old Salt, and, best of all, he welcomed me aboard with unrestrained warmth. Perhaps it was because I was the only person that he'd ever met who never tired of his monologues.

One day, a local yachtsman moored nearby told me an amusing story. A small speed-boat had appeared in the creek, buzzing the yachts, and Harry came up on deck, naked, and began yelling abuse. Then he leapt into his dinghy and rowed furiously over to the yachtsman's boat. He talked animatedly to the man and his wife for some time, as was his wont, before glancing down. 'Oh my God,' he exclaimed, red-faced, 'I'm naked.' He rowed back to *Kelasa* as fast as he could, leapt into his cabin and slammed the hatch, as if the devil was after him.

With a straight face, the man telling the story said, 'My wife was very impressed. I felt somewhat inadequate that night when we undressed for bed.'

Harry told me the speedboat story as well, omitting to mention that he was naked. 'I was hoping they'd run into someone's mooring lines and garrote themselves,' he said.

*Although posed for the press, this photo shows something of the complex relationship between Adrienne and Harry. Photo: courtesy David Matzenik.*

When Adrienne and Harry met, they spent a few months sailing *Kelasa* on Sydney Harbour, sometimes with Adrienne's boat-mad 17-year-old son, David, aboard. David came to Durban to join *Kelasa* in 1970, aged 19, but left on another yacht because Harry had been particularly difficult with him.

*Kelasa sailing on Sydney Harbour with 17-year-old David Matzenik on the helm. Photo: courtesy David Matzenik.*

Once, when David criticised Harry to Ken Furley, skipper of the Australian yacht, *Fortuna*, Ken said, 'The thing you need to appreciate about Harry is that he got out from under.' And that he did. Harry lived on *Kelasa* for the next 30 years, and sailed the boat around 100,000 nm, two-thirds of which were singlehanded. Apart from the manual anchor winch, all gear was handled with purchases or tackles, including the tiller in rough seas.

Self-steering was achieved by taking the staysail sheet to the tiller. Harry had refined this system, by fitting tracks to both sides of the tiller, to easily adjust the position that the sheet was attached, and also had tracks for the turning blocks on either side of the boom gallows (see photo of Adrienne farewelling Sydney above). For light winds, the sheet was moved towards the inboard end of the tiller, coming aft towards the rudder-head as the wind increased in strength. Rubber shock-cords on the opposite side of the tiller balanced the forces.

*Kelasa's* ancient petrol engine was a character in itself. Harry had rescued it from a paddock, where it had lain for many years, rebuilt it and called it *Vivienne*. It had exposed valve springs and rockers, if I remember correctly, and was notoriously unreliable.

In Honolulu, a few years later, he replaced it with a slow-revving, single-cylinder 15hp Yanmar diesel, which had a huge flywheel. Designed for fishing boats, it was not normally available to the recreational market, and Harry got it from a Japanese-Hawaiian fisherman. It swung a large propeller and was more reliable than *Vivienne,* but significantly underpowered for a 22-ton vessel. As far as I know, it didn't have a name, unless it was one of the swear words Harry used when swinging that huge flywheel, since it had no electric starter motor. As always, *Kelasa* remained without battery or electricity.

*Fortuna* was a carvel-planked cutter, heavily built with grown timbers in Hobart, Australia, in 1944. It was built from the half-model of a yacht called *Chloe,* which had been produced by Percy Coverdale, a Tasmanian boatbuilder of some repute. He was famed for building his boats from hand-carved half-models, rather than paper designs, and they had a reputation for being fast and seaworthy. Built by its first owner under Percy Coverdale's supervision, *Fortuna* had sailed in seven Sydney-Hobart races, the first in 1947, and came close to winning the race a couple of times.

Ken, a New Zealander, stayed in Durban for a year. Being one of the larger voyaging yachts (these days it would be considered small), it was moored against the jetty, with several others rafted outboard. I loved having all these visiting yachts rafted together. For a solitary boy who'd lost his heart to the lure of the tropical trade winds, salty boats and faraway places, it was a delirious dream. I remember my shock when Ken growled one day, 'I can't wait to get the f..k out of this place.'

*Fortuna alongside the International Jetty with several yachts rafted alongside. These days I can better appreciate Ken's frustration with people walking across his decks day and night. Photo: courtesy David Matzenik.*

Rodney Perkins, by way of contrast to Harry, could be described as a comedian, originally from Wales. In his mid-thirties, with a well-trimmed brown beard, he claimed, tongue in cheek, to be a remittance man.

He built his 32' carvel-planked ketch, *Weir*, over a five-year period in Sydney, Australia. *Weir* had a white hull, lots of varnish and an elegant finish. It was a solid boat, but not as solid as *Kelasa*. Being moored alongside *Kelasa* made *Weir* look almost dainty. The boat was a sistership to William Albert Robinson's *Svaap*, another early circumnavigator, though Rodney did not like to say so, as he bought the plans from a designer in Sydney who had been inspired by *Svaap*, in much the same way that the plans of *Suhaili* were inspired by Atkins' *Eric*. Both designs are exact copies of the yachts that 'inspired' them.

*Rodney Perkins holding court on the International Jetty.*
*Photo: courtesy David Matzenik.*

*Weir sailing on Sydney Harbour prior to departing for Wales via South Africa. Photo: courtesy David Matzenik.*

*Weir's* reputation preceded it into Durban. The crews of other yachts who had met them in Chagos Archipelago and the Seychelles were talking about the crew and their amusing habits. They addressed each other formally, and encouraged others to do so too.

Thus, Rodney was always Mr Perkins, and his crew, when *Weir* arrived in Durban, were Mr Butterworth and Mr Conway Hyphenated Physics.

Being a qualified engineer, Mr Perkins had no difficulty procuring work, but nothing lasted. He always got sacked, or *expelled* as he preferred to put it. It wasn't that he was incompetent, far from it, but his sense of humour kept getting him into trouble.

'Not enough forelock tugging on my behalf,' he said, 'and too much of the other sort on theirs.'

On one occasion, for instance, the supervisor had been an obnoxious character, obsessed with the idea that his crew were as intent on lying, cheating and stealing as he had been before his elevation. One morning, a few minutes after start-up time, Mr Perkins realised that he'd left an important tool in his locker, so went back into the tea-room, where the lockers were placed, to get it.

He'd just entered the room when the supervisor sprang in behind him, shouting, 'Aha, so what are you doing in here, hey, at this time of the day?'

'I'm waiting for morning tea,' said Mr Perkins, sitting down casually at a table. And that was the end of another promising career.

Once, I overheard a conversation between Mr Perkins and Adrienne, whom he called Mrs M. 'What you need, Mr Perkins, is a good woman,' said Adrienne, perhaps eyeing the post herself, for Mr Perkins was a youngish and comely man.

'Oh, I tried sex once, Mrs M, but it made me dizzy,' Mr Perkins replied, poker-faced.

When Rodney eventually sailed from Durban, bound for England, he took David Matzenik with him as mate on *Weir*. It was a formative experience for David, even though he'd been keenly interested in voyaging under sail since he was a boy. Rodney was generous with his time and knowledge, and helped David evolve into a fine seaman. The voyage became a catalyst for David's long career as a professional yacht captain, and he and Rodney have stayed in touch through the years. Rodney kept *Weir* for many years, and has only recently sold the boat due to advancing age.

In the Azores Islands, *Weir* crossed paths with a famous English yacht, *Tern IV*, once owned by Claude Worth, who had mentored Eric Hiscock and wrote books considered the last word on seamanship in his day. On reaching England, David joined *Tern IV*, assisting in the restoration of the vessel's original gaff rig. *Tern IV's* owner, Roger Fothergill, was one of the first people in the Caribbean charter trade to restore a classic yacht to its original rig, something that later became a cultural trend.

*Weir alongside the International Jetty.*

The following winter, David sailed out to the Caribbean aboard *Tern IV*, and rose up to serve as mate on numerous charter vessels, including the replica of *America*, and the famous *Ticonderoga*, before taking command of boats like *Puritan*, the 103' Alden schooner, and the Fife classic, *Altair*. I was in awe of David when I met him in Durban. He seemed so debonair, handsome and self-confident, and I envied him having a mother like Adrienne. He became another lifelong and much-cherished friend.

One yacht that sticks in my memory from that summer is *Myonie*, sailed by a middle-aged American couple from Miami, Al and Helen Gehrman. *Myonie* was a 36' Carol Ketch, big sister to John Hanna's famous and ubiquitous Tahiti Ketch, which had been based on a Greek sponge-fishing boat.

*Myonie* was carvel-planked and gaff-rigged, with no fancy gear, no spray dodger, cockpit shelter, or self-steering mechanism, though it did have a powerful diesel engine. They used the staysail sheet-to-tiller system for self-steering when conditions were suitable, or hand-steered. The hull was matt-white, and Al had painted the trim brown. 'Almost as good as varnish from a distance,' he claimed, giving me his larrikin grin. There was no gloss on *Myonie,* or Al.

They were on their second round-the-world voyage, the boat on its third. After the first, Al had abandoned his career as a doctor, when he discovered he could make a good living *dumpster-diving*, hauling perfectly good gear out of commercial dumpsters around yacht clubs and marinas. Often the salvaged goods only needed cleaning or re-wiring. Al would fix the equipment, give it a bit of a polish, and sell it at the week-end markets.

*Myonie on Wilson's Slipway, December 1969.*

'It took four years to save up for the first circumnavigation,' he said, 'but only eighteen months for the second. I didn't have to pay medical insurance, secretaries' salaries, dry-cleaning bills, or surgery rentals. Dumpster-diving is an undervalued profession.'

I looked at him askance, not sure if he was pulling my leg, something he loved doing.

'You may think it's disgusting,' he laughed, with a gleeful expression on his face, misinterpreting my response, 'diving into garbage bins alongside the feral cats and old winos looking for a feed, but, hell, I bet you've never had to stick your finger up some old guy's bum.' And he laughed, his wide, weather-beaten face creased in merriment. It was hard to imagine this sun-burned, irreverent man wearing a white jacket in a doctor's clinic, but he may well have been a better doctor than most.

Al and Helen completed four circumnavigations aboard *Myonie* eventually. Nothing seemed to faze Al. After leaving Darwin in 1969, they'd broken the bowsprit in a gale south of the Cocos-Keeling Islands, meaning that they could not use their staysail sheet-to-tiller self-steering arrangement, and had to hand steer for the next 5000 miles non-stop to Durban, because they were late in the season and needed to get out of the cyclone belt. Al just laughed it off.

Instead of giving straight-forward interviews to the newspapers, he read them a poem he'd written. I've forgotten most of it, apart from the lines: *And so we came to Australia / Where the natives sure talk funny / We didn't see any kangaroos / But we sure spent lots of money.*

During their fourth circumnavigation, they came into Suva, Fiji, late at night, and anchored in the main roadstead. Being exhausted, they forgot to put up a riding light and were almost run down by a coastal barge returning to port. The barge skipper published a scathing letter in *Pacific Islands Monthly*, imploring yachtsmen to be more responsible, and he also made an official complaint to the Suva Port Captain. I am sure Al just shrugged, and gave the irate port captain his larrikin grin.

*Youth in Nukualofa, Tonga, 1966. Photo: courtesy Rags Weldon.*

Alan Quigley, another thirty-something, hailed from Adelaide, Australia, and sailed in with an all-male crew on the 36' steel-hulled sloop, *Youth*, a typical cruiser-racer from the late 1950s. At the tender age of 23, he'd built the boat himself, to a design by West Australian, Len Randell. With his crew, according to legend, he had barnstormed around the South Pacific, establishing intimate relationships with various coral reefs and the unattached daughters of colonial administrators.

*Returning to Youth at Palmerston Atoll, Cook Islands. Rags Weldon centre, playing the guitar, Bill Legler, left, Max Graham, right.*

*Youth aground in Maupiha, 1967. Alan Quigley on the end of the boom, then Jim Mahoup and Trevor Bone. They are trying to heel the boat and reduce draft. Photo: courtesy Rags Weldon.*

There was only one, unbreakable, rule on *Youth,* as explained to me by Rags Weldon, one of Alan's crew in Durban, *no fighting in the yacht club.* It eloquently framed their reputation, deserved or otherwise, of being a tough bunch of charmers. I was scared of Alan and his crew, yet I had the distinct impression that Alan would risk his life to help you if the need arose. Years later, I met him again, and have been in contact in recent years with Rags Weldon on social media, and they both seemed quite tame. Did I just imagine how wild they were? I was, of course, very young, and shockingly naive.

Quigley, as everybody called him, always had a gang around him, a non-stop party where rum flowed, jokes flew, and laughter reverberated. When *Youth* eventually sailed from Durban, there were several new hopefuls on board, who'd all willingly paid a large sum of money to join the Quigley Club, on top of daily expenses.

I was standing on the jetty one day when a small, red-hulled, sloop sailed in. John Sowden, sailing alone aboard 25' *Tarmin,* was 21 days out of Mauritius.

*Tarmin in Durban, 1969, alongside the Vertue, Sekyd, which sailed out from England via the Red Sea. The large American trimaran, Ellimata, lies astern.*

An American, John had lived in Spain for a number of years, and was something of a magician, having performed on stage in Sydney, Australia, shortly after World War Two. He was 49 years old, but still had a youthful demeanour and an excellent, casual dress code. Occasionally, at parties, he could be induced to demonstrate a few magic tricks, such as losing his cigarette, and then, after looking for it everywhere, and asking everybody around him to look for it also, often to the consternation of his host, taking it out of his mouth.

Magic was just a hobby, although there was a playful quality in everything John did. Many years later, I discovered that he'd recently retired from an international career installing radar systems in airports, etc. Perhaps having such an intimate knowledge of complex electronics influenced his decision to voyage in one of the smallest, simplest boats imaginable. Apart from a wind-driven self-steering mechanism, *Tarmin* was entirely devoid of special equipment. John didn't even have a spray dodger over the main hatch, or an awning. Later, he hand-stitched himself an awning.

John bought the carvel-planked *Tarmin*, a Yachting World 5-Tonner designed by English designer, Robert Clark, in Mallorca in 1966. The boat had been built in 1949 in England and was registered in London. It was in a decrepit state, and he saw it as a winter project, not intending to do any serious sailing. But after patching up the old boat, he'd drifted from Spain to the Canary Islands, thinking the climate there might be more congenial for working outdoors in winter. In the Canary Islands, he realised it was easier to sail on to the West Indies than beat back against the prevailing winds to Europe.

*Tarmin on Wilson's slipway, preparing for the last leg of John Sowden's circumnavigation. Note the bowsprit has been refitted. John took all the metal deck fittings off in Durban and had them regalvanised.*

'After that,' he said, 'it just made sense to keep going.' Now here he was in Durban, still working on the boat. It may have been decrepit once, but he kept it immaculate now. Its bright red hull gleamed, its mild-steel metal fittings were painted silver, and he was always working on something or other.

The only thing that seemed to defeat him was the Stuart-Turner petrol engine, perhaps because it lived under the non-self-draining cockpit that emptied into the bilge. John covered the engine with a sheet of plastic when at sea. Numerous attempts had failed to keep the motor running, and for much of the 20 years that John roamed the oceans, *Tarmin* sailed engineless.

'You don't need one for this sort of sailing,' he said, 'though it takes some of the worry out of entering and leaving port.'

I looked dubiously at the large, open cockpit. 'What happens when waves break into it?' I asked.

'Oh,' said John with a quiet smile, 'the water just goes into the bilges and I pump it out again. It's safer than having all that water trapped at the back of the boat when the next wave rears its ugly head. Besides, it keeps the bilges clean.'

He had a calm, simple approach to seamanship. He never entered port at night, preferring to heave-to offshore under triple reefed mainsail and wait for daylight. Patience, he said, was a seaman's greatest virtue.

*John Sowden topping up with water before departure, January 1970.*

'Are you going to write a book?', I asked.

'I couldn't,' he said, 'there would be nothing to put in it. Nothing ever happens to me.'

On the passage from Durban around the Cape of Good Hope, a ferocious SE storm sprang upon the ship as they approached Table Bay, just a few miles short of Cape Town harbour. John had already poured himself a congratulatory glass of whisky. He ran *Tarmin* off downwind, NW into the Atlantic, with the storm jib sheeted flat, fore and aft, and every spare rope in the boat, plus the bosun's chair, streaming out behind. The wind blew at over 100 knots for 18 hours in Cape Town, according to Croix Grut, who was there in *Iorana*.

Unable to keep his oil lamps burning, he lashed an electric torch to the rail, illuminating the sail, and hoped for the best. *Tarmin* was in one of the busiest shipping lanes in the world, and visibility was almost zero. Then he went to bed and slept through the night, as best he could amid the howling cacophony. When he woke, he found 200mm of water slopping over the cabin sole. It took *Tarmin* a week to beat back to Cape Town.

*Tarmin under sail. Photo: courtesy Joan Sowden-Deily.*

He was one of the happiest lone sailors I ever met, content with his own company yet willing to share a joke or a cup of tea when the occasion presented itself. He claimed to be a natural-born loner. Between 1966 and 1986, he sailed *Tarmin* three times around the world alone, earning himself a place in the Guinness Book of Records, which pleased him, but he had no desire to seek further recognition. He sailed for his own pleasure.

*Tarmin* made a number of very long passages, twice being at sea for 90 days or more, once from St Lucia in the Caribbean to Recife in Brazil, en route to Rio de Janeiro, and once from Panama to San Diego. *Tarmin* had to beat against prevailing wind and currents on both occasions, and these passages are considered notoriously difficult for small, engineless craft.

On the trip down to Brazil in October 1972, with *Tarmin* pitching heavily into steep head seas, John fell while changing headsails on the foredeck, landing on the pulpit and breaking a rib. For some weeks afterwards, he had to sleep sitting up, wedged in a corner of the leeward bunk, until his rib healed.

*Tarmin in St Lucia, West Indies, at the end of John's third solo circumnavigation in 1986. This photo was taken the day after he arrived from Durban, with just one brief stop at the open roadstead in Saint Helena Island. Photo: Herb Smith.*

He chose not to stop in Cape Town during his second and third circumnavigations, sailing non-stop from Durban to St. Helena Island, which has a very exposed anchorage where rest is next to impossible, then on to the Caribbean. Cape Town Harbour, as many voyagers have found, was very dirty in those days, with crude oil in the harbour soiling lines and topsides, plus black soot on deck from the steam locomotives that shunted back and forth behind the yacht club.

John died from bowel cancer in 1990, four years after the end of his last circumnavigation, and just short of his 70$^{th}$ birthday. His story has been recorded for history by his youngest sister, Joan Sowden Deily, in her book, *Alone on the Sea*, compiled from John's extensive logs, tape recordings and letters.

There was an audible gasp one Sunday morning when the biggest trimaran in living memory motored in, flying the American flag. *Ellimata* looked like an aircraft carrier. It was a 45' Hedley Nicol Voyager design, built in cold-moulded plywood in Brisbane, Australia, with a 24' beam and accommodation in all three hulls. It looked like Tom Corkill's little trimaran on steroids.

In the first couple of hours after *Ellimata* tied up, dozens of people traipsed across the decks. I was almost the first person aboard, and was still there late in the afternoon when most people had left. The skipper, 45-year-old Curtis de Camp, wanted to go for a walk and look at the town. I went with him, to show him where the post office and other important venues were. The streets, late on a Sunday afternoon, were almost empty.

'Where are all the black people,' Curt asked, 'I thought this was Africa?'

I had to explain that the Pass Laws kept them out of the city after commercial hours, unless they were domestic servants, and most of those were in the suburbs. Curt was reeling as he walked, trying to find his land legs. A muscular, deeply-tanned man, he was bare-chested and barefoot, clad only in a pair of tattered denim shorts.

He had been a successful businessman in some mid-west American city, but had tired of being held up at gun-point by drug-addicted criminals, and always packing his own gun for personal safety. He'd sold up, moved to Brisbane, Australia, and built the trimaran. With a luxurious interior to accommodate guests, *Ellimata* was now headed for the West Indies to begin a career in the charter industry. I was excited to hear that Tom Corkill had been in the same boatyard, building a catamaran called *Ninetails*.

During our long walk, Curt, who had two daughters, said, 'I wish you were my son. I'd take you back to the States and we could go fishing and hunting together.' I did not tell him that I loathed fishing and hunting, and had a passion for ballet.

*Ellimata careened on the sandbanks inside the Durban yacht basin for a quick scrub and a coat of antifouling paint.*

From that day on, I spent most of my time aboard *Ellimata*, helping out with maintenance, listening to Curt's stories, and enjoying the feeling of being a part of the voyaging community. The other voyagers often dropped by for a yarn and a cup of tea. That summer felt like my debutante ball, like *My Own Private Idaho*. They were happy days, when I was able to put aside my sense of alienation, my disappointment of not sailing to New Zealand with Clive, and the looming, much dreaded, prospect of National Service.

I missed *Ellimata's* departure for Cape Town, although I cannot remember why, perhaps I could not bear it. I do remember coming down to the jetty the following day. There seemed to be a great big hole in the water where *Ellimata* had been. Or was it in my heart?

# Chapter Six

# Between a Dream and the Reality...

Books have always had an irresistible appeal for me. I've bought thousands of them over the years, often spending money I could ill afford. I purchased titles on ocean cruising, philosophy, psychology and art, but my interests were always esoteric; I was searching for adventure, freedom, beauty and truth. I gave most of them away, for various reasons, but kept the voyaging titles.

Twice, people who I considered friends stole a large number of my sailing books, and I purchased them all over again. My boats have always been small, and half the lockers aboard have been stuffed with books. I could have lived for several months each year on the money I spent. So much for the theory of buying the smallest boat I could squeeze two bunks into, working for three months a year, then stocking up with beans and rice, and sailing off again into the blue yonder.

By the beginning of 1970, I'd already purchased twenty-five ocean cruising books. Being such a good customer, I'd become friendly with a woman in the best bookshop in town. She would put aside the latest titles for me, or order one in if she saw it in her catalogue. I'd walk into the store and she'd produce a new title with a little flourish. I think she enjoyed my delight.

One day she mentioned that relatives of hers, a young couple called Karen and John Cross, had just purchased *Vertue Carina* and were preparing the boat for a voyage to Denmark, via England, to visit the birthplace of Karen's grandfather. She introduced us, and I started visiting John and Karen aboard *Vertue Carina.* The sweet little boat was still moored fore-and-aft on the trots in the yacht basin.

It was thrilling to finally step aboard a Vertue, and *Vertue Carina* at that. The boat seemed so seaworthy, with its stout Burmese teak coamings, solid-teak locker doors, and extra Perspex bolted to the outside of the windows. You could see at a glance that this was a deep-sea boat. Bruce Dalling, the boat's original skipper, had become one of the iconic characters in my pantheon of sailing heroes, all the more intriguing since he had not written any articles or books. The association with Dalling added to *Vertue Carina's* lustre in my mind, not that it needed it.

I was convinced that if I could just get my hands on a Vertue, I'd be invincible, and my doubts, fears and distress would evaporate. But how could I ever afford such a magnificent yacht? John and Karen, 25 and 24 years old respectively, were both university graduates, earning professional salaries, and had been saving to buy their boat for several years. I didn't want to wait that long. Patience was not my long suit.

*The interior of the Vertue-class yacht, Fidelis.*
*Photo: Wooden Ships Yacht Brokers.*

People in the yacht club told me that the Vertue was too small for ocean cruising, or even just for living aboard. These naysayers were mostly middle-aged men, and I looked disparagingly at their well-fed bellies, telling myself that I was never going to be that fat. I'd always be lithe, able to wriggle in anywhere. Oh, the sweet arrogance of youth. Not that I'd need to wriggle, I thought; *Vertue Carina* seemed more than spacious enough.

I still like Vertues, and could certainly live on one, though they do seem a wee bit tight to me today. There are two narrow berths amidships, good sea-berths, with alcoves to stick your feet into. This means that the saloon, where you sit on the bunks, is 4' fore and aft and 6' wide. All well and good. However, I would prefer not to wriggle up under the cockpit these days.

By the time I met Karen and John, I was working on my third job. I lasted two weeks as a trainee bank clerk. It was *soooo* boring, Clive was right. That was followed by less than one week as a cadet deck-officer on a coastal cargo ship. That should have been ideal, as the First Mate was a sailing man, who'd sailed from Cape Town to the Caribbean on one of Arthur Holgate's schooners, but he gave me a particularly hard time. 'Our yachtsman', he called me, perhaps because the first time I appeared on his deck I was wearing flip-flops, or beach sandals. I couldn't take the discipline. When he threw his dirty boots at me to clean, it was all over.

Now I was working as a trainee auctioneer in the fresh produce markets run by Durban Corporation. Up in the office in the afternoons, it was as boring as the bank had been, but the produce floor, where I worked in the mornings, was wild enough to entertain me, with some eccentric characters among the growers' agents. There was nothing genteel on the produce floor.

During lunch breaks, instead of sitting down and eating, I used to run down to the yacht mole to see if any new yachts had arrived, or to talk to the sailors I'd befriended. Once I went and visited *Clipper*, which was up on Wilson's Slipway, where Clive was getting the boat ready for his second attempt at sailing to the Seychelles. I was sad I was not sailing with Clive, and *Clipper* looked so beautiful up on the slipway cradle that it made my heart ache in a way I couldn't get enough of.

I had settled down in the markets, but the wages were low, and I realised that it would take a very long time to save up the amount of money John and Karen had paid for *Vertue Carina*. My boat would have to be smaller and cheaper. There was another Vertue for sale in Durban at that time, *Aldebaran,* built locally by Henry Vink and sailed to Mauritius. I think he knew I was a dreamer, but he let me crawl all over the boat, and spent a couple of hours chatting with me in *Aldebaran's* cockpit.

He was asking ZAR 5000. A South African rand in those days was equivalent to an American dollar. I desperately wanted *Aldebaran*, especially since the boat had featured briefly in the *Sandefjord* movie, when it was seen sailing out of Port Louis, Mauritius, but I might as well have wanted the moon. *Aldebaran* later completed a circumnavigation of the world and is now based in Cape Town.

*Clipper on Wilson's Slipway prior to Clive's departure for the Seychelles. Photo: Clive Rouse.*

I really should have taken a 'tack offshore', gained some qualifications that would have earned me better wages, and then started saving for my boat. Unfortunately, I was wedded to the idea that I could find a small, cheap boat to fix up quickly, and then wash dishes for three months a year to pay for the baked beans. Plus, there was no way I was going to stay in South Africa. This geopolitical reality, and my psychological blindness, would cost me dearly.

John and Karen had little sailing experience, but they set off for Cape Town and survived a difficult passage, which included John having to go overboard in a gale off Port Elizabeth to free a rope that had wrapped around the propeller. From Cape Town they had a good run to Saint Helena Island, 1700 miles distant, though they were depressed and worried at times, especially about whether they would be able to find the island, given that this was their first attempt at celestial navigation, and also because there were many cloudy days when John couldn't take sun sights with his sextant. Saint Helena is only five by ten miles wide, and even experienced navigators sometimes had difficulty finding it in inclement weather.

John and Karen also had difficulty adjusting to the motion, and were often seasick. Vertues, like all slack-bilged English designs of that era, roll heavily in the open sea when the wind is aft of the beam, although, paradoxically, it is their moderate beam, deep draft and slack bilges that give them their enviable seaworthiness.

Arriving at Saint Helena Island, 17 days after leaving Cape Town, seemed to turn the tide for them psychologically. They said that there was never another place they felt quite the same about as St Helena, due in part, no doubt, to the euphoria they felt after their landfall. From there, they voyaged confidently and joyfully onward to Denmark, via Ascension Island, the Azores Islands and Falmouth. They were at sea for 56 days en route from Ascension Island to Fayal in the Azores Islands.

After visiting Denmark, they came back to Falmouth for the winter. The following summer, Karen gave birth to a son, Peter, and when the boy was four months old, they set sail via Portugal and Spain to Brazil. From Rio de Janeiro, they sailed back to Cape Town across the South Atlantic Ocean, a stormy, cold passage that lasted 53 days.

*John, Karen, and baby Peter Cross, shortly after their arrival back in Cape Town. Photo: courtesy Karen Cross.*

After 27,000 miles, three years and one day, they had brought *Vertue Carina* home, completing a faultless three-year voyage. That last passage in particular, skirting the roaring forties of the Southern Ocean as autumn approached, looking after the boat and a baby who was becoming more active every day, is one of the great Vertue stories.

Today, John and Karen live in Simonstown, east of Cape Town, and *Vertue Carina* is moored nearby, though they no longer own the boat. They have a modern racer-cruiser now that John races locally. Karen wrote an inspiring book, *The Baby Boat*, about their adventures.

There was a 20' canoe-sterned, strip-planked yacht in Durban that took my eye, designed and built by Peter Strong, who ran Wilson's Slipway. It had a lovely hull shape, something like a miniature *Tzu Hang,* and was on the market for ZAR 600. It was out of the water for a while next to the International Jetty, in the same spot where *Ohra* had been rebuilt a few years earlier, and John Sowden looked it over with me. He said it would be a good boat for a young guy like me (able to wriggle in anywhere), but I didn't have the money, as little as it was.

In those days, no bank was going to lend a boy money for a boat, even if he was in stable employment. I can see now that I would have needed to spend a lot more money to get it ready for ocean cruising, but it might have given me a good base to start from. I sometimes dream of what might have been if I had managed to secure it, fit it with an unstayed mast and junk sail, a large scull to get around when there was no wind, plus some sort of self-steering gear, and sailed off to the Caribbean. Mind you, there wouldn't have been any space for my books.

Peter Strong was something of a local identity around the Durban waterfront. He had once employed Bernard Moitessier, against his better judgement, during the latter's lay-over in *Marie Therese II*. In Moitessier's book, *Sailing to the Reefs,* there are several amusing anecdotes about Peter. He was a talented designer and shipwright, who had designed and built a number of successful boats, including one last one for himself at the age of 71, the 50' carvel-planked ketch, *Wayfarer*. He designed *Wayfarer* as a downwind flyer, and campaigned the boat in the 1971 Cape to Rio yacht race.

Despite starting a day late due to a slow passage from Durban against headwinds, *Wayfarer* completed the race in 29 days and 17 hours, beating half the fleet. The boat might have done even better, except for a spectacular broach one day that put the masts in the water. The crew weren't too keen to set spinnakers after that, according to Peter, and they all left the boat after the race. Peter had hoped to sell the boat in Rio, but, failing to find a buyer, sailed back to Cape Town with two inexperienced Dutch boys as his only crew. He spent much of the passage on his hands and knees, changing headsails on the foredeck. It was a slow passage, and towards the end of the trip they had to break into the liferaft to access its emergency rations.

I decided that since I couldn't raise money to buy a boat, I'd build one, reasoning that I could purchase one plank at a time. It was folly, in hindsight, because new boats always cost more in the end, and I was not a practical youth. I'd never used tools, had no patience, and had, I later realised, some sort of learning difficulty when it came to technical matters. It might have been wiser to stick to my original idea, which was to save up a couple of thousand rands, go to England, and buy a clinker-planked timber Folkboat. They were being advertised in the English yachting magazines at that time for around £1000. Of course, I was not budgeting for the ancillary costs. Dreamers never do.

My father was a talented craftsman who could turn his hand to anything, from mechanics to carpentry to bricklaying, but he said he was not going to help me build a boat so I could go out and drown myself. He wasn't going to help me buy one either, for the same reason. It is a pity, because building a boat together could have been a good bonding experience, and I would have ended up with a yacht far less likely to drown me than anything I could produce on my own.

What boat to build was the question. I was thinking, of course, of a small, simple, cheap boat that I could knock up quickly. John Guzzwell's *Trekka* seemed the perfect size, even smaller than *Dove*, but I was worried about the complexity of the design. I considered hard-chine plywood designs, like those produced by Robert Tucker, before I fell in love with Paul Erling Johnson's strip-planked 28' gaff ketch, *Venus II*.

I already knew about Paul, or Johnson, as his friends called him, having found an article in an old *Practical Boat Owner* magazine about his first *Venus*. In the 1960s, that publication was still the platform for small, funky, one-off boat projects. An older friend of mine, Lawrie Williams, had a huge stack of these magazines that I loved to dive into whenever I visited him and his wife, Lorna. They'd invite me for dinner and I'd spend most of the evening with my head in a magazine. Lawrie and I would then have long, animated discussions about the various boats I came across, before he took me home across the city on the back of his motorbike.

Lawrie and Lorna owned a 24', 5-ton gaff sloop, *Lapwing*, of some vintage, that was one of Frank Robb's original Royal Cape One Design Class (RCOD). Frank Robb was the author of what was once considered an authoritative work on storm tactics, *Handling Small Boats in Heavy Weather.*

The original RCOD was eventually replaced by a lightweight, single-chine, 30' Van der Stadt plywood sloop, based on the Black Soo design. Frank Robb declared that the new design would be a death trap in the first serious gale it encountered, but when it not only didn't break up, but trounced every other yacht in the fleet, he decided to give yacht design away.

I coveted *Lapwing* but Lawrie and Lorna wouldn't sell it to me, as they considered it too old and tired. Another young couple eventually bought the boat and sailed it to Cape Town after a refit. Lorna has long since passed away, but Lawrie and I still regularly talk on the phone.

*Little Venus*, as it came to be known, was originally an 18' Shetland Island fouraareen, an open-decked, clinker-planked, double-ender used for fishing. It had been built from driftwood in 1898 on the island of Foula. As a young man in the early 1960s, having just completed his national service, Johnson decked *Venus*, rigged it as a gaff ketch, then sailed solo to Norway, the Canary Islands, the Caribbean and Florida, without engine or electrics. Johnson grew up afloat, on his parent's Colin Archer ketch, *Escape*, and he was a natural seaman.

Later, I found a reference to *little Venus* in Bernard Moitessier's book, *Cape Horn the Logical Route.* Moitessier met Johnson in late 1964, in Las Palmas, Canary Islands, and published a photo of *little Venus* sailing past *Joshua*. That photo had a profound influence on me. It re-enforced my belief that all I needed was the smallest boat I could shoehorn myself into. Moitessier was very impressed with Johnson, describing him as a first-class seaman and a tough mariner. As the old salts say, it's not the ships but the men, or women, in them.

After crossing the Atlantic to the Caribbean, and cruising up the Intracoastal Waterway, Paul replaced little *Venus* with a new boat, the 28', semi-flush-decked *Venus* II, another double-ended, engineless, gaff ketch. He sketched his ideas for a small, Colin Archer inspired double-ender, had the English yacht designer Percy Dalton draw up the lines for him, and built the boat in strip-planked timber he salvaged from an old church in Florida. The bolt-on keel was cast in re-enforced concrete, and he sewed his own sails.

The new boat cost him less than US$1000, and he spent years crisscrossing the North Atlantic in it.

*Little Venus sailing past Bernard Moitessier's Joshua in Las Palmas in October 1964. Photo: Bernard Moitessier.*

*Venus II under sail. Photo: venusketch.com*

*Venus* II was also written up in *Practical Boat Owner* and seemed ideal for someone with a limited budget. At least Paul made it sound that way. But how many people are able to find an abandoned church with enough seasoned pitch pine lying around to build a whole boat? Even if it did have a concrete keel, home-made sails and no engine, it was still a lot of boat for the money.

It was a much larger boat than I had intended to build, but I was seduced by Paul's costings. I wrote to Paul, care of the magazine, but received no reply. Years later, I heard from friends of his that he would probably have given me the plans if I'd met him, but he was not a good correspondent, being dyslexic.

Later, little *Venus,* the 18-footer, ended up in the Scottish Maritime Museum, restored to its original form. I thought it was a pity the museum stripped the boat back. I feel sure there are other examples of Shetland Island fourareens, but there was only one *Little Venus.*

I decided, therefore, to have another look at James Wharram's catamarans. They had also received write-ups in *Practical Boat Owner*, and the magazine regularly displayed advertisements from the designer, with evocative sketches of various models. They were the cheapest, simplest boats to build, according to James Wharram, in which people could escape the soul-destroying cities and find freedom on the ocean. Anybody could bang one up in the backyard, he claimed. It sounded good to me.

My interest in small multihulls had also been rekindled by Tom Corkill, who had recently arrived in Durban on his new boat, the 34' catamaran, *Ninetails.*

*Ninetails under construction in Brisbane, 1969. Photo: Tom Corkill.*

*Ninetails cruising up the Queensland coast.*
*Photo: Alan Lucas.*

Other yachtsmen suggested that Tom should have called the new boat *Nine Lives*. Like his little trimaran, it was very basic. It did not seem to be quite finished, had no radio, only one set of sails, and very few charts. At least this time Tom had a plastic sextant.

He seemed unfazed as usual, although he was not that happy with *Ninetails* when he arrived in Durban, complaining that the designer he'd commissioned had just taken a set of line drawings for the 30' Iroquois catamaran, published in *Yachting World* magazine, and bumped 4' onto the back of it.

Getting frustrated while trying to fair the hulls during construction, he'd sneaked a look in the designer's files one day and found the evidence. I asked him which of the two boats he preferred, the new catamaran or his little trimaran, and he quickly said *Clipper*.

After cruising up the coast of Queensland, Tom left Darwin in May 1970 with a young woman, Patricia Reinsma, as his sole crew, and sailed to Durban via Indonesia, Mauritius and Reunion over the next five months. Their boisterous passage from Bali to Mauritius took 23 days, they suffered a broken rudder and ran short of water. En route from Mauritius to Durban, with two extra crew aboard, they survived a severe storm that capsized a tanker off the coast of South Africa. Patricia left the boat the day they arrived in Durban.

Tom soon took off for Cape Town and on to Rio de Janeiro, the port he'd been heading for when *Clipper* capsized. En route to Rio, he sailed past the spot where the little trimaran had capsized almost three years to the day it happened. Considering that he lost everything he owned when *Clipper* capsized, it was a remarkable turnaround.

The ship that rescued him had taken him to India, where he worked for a while as an outreach officer for a religious organization, then headed to Vietnam where he managed entertainment programs for American troops for a year.

He eventually sailed *Ninetails* two and a half times around the world, as well as making numerous passages between Australia and Asia. He also delivered yachts all over the world. In Australia, he became a bit of a legend in multihull circles, but because he never publicised his voyages, is largely unknown elsewhere. He was one of those sailors who gave me the appearance of being fearless, despite being so quiet and unassuming.

*Tom and Sherri Corkill aboard Ninetails. Photo: courtesy Tom Corkill.*

The early Wharram catamarans, now called the Classic designs, had a small cabin in each hull and a slatted deck in between, inspired by the twin-hulled voyaging canoes of ancient Polynesia. James Wharram had sailed across the Atlantic Ocean four times, as chronicled in his book, *Two Girls, Two Catamarans,* published in 1969.

After reading his book, I wrote to James Wharram and received a quick reply. For the next few years, I corresponded with both him and his partner, Ruth. In one letter, Ruth told me that I reminded her of James when he was my age. Full of energy and social outrage, I think. I decided to build one of James's 27' Tane designs, though I think the 34' Tangaroa would have been better suited for my liveaboard voyaging aspirations.

I started out with such high hopes. The day that the plans arrived, in a large, fat package, was better than an orgasm. I rushed out and bought a few sheets of plywood, then started madly sawing them up. I had no power tools but it didn't matter. I was 19 and had energy to burn; I was on fire. My father stayed clear. Only once, watching me put together a dubious joint, glue everywhere, all over my hands, my shirt, in my hair, he said, 'Maybe you should build three hulls and throw the first one away.'

*The hulls set up behind my parent's house for assembly. Everything, as you can see, was only half-finished.*

It was good advice, except for the fact that the third hull wouldn't have been any better than the first. I had no experience with tools, plus I was in too much of a hurry to learn. That takes time, and the average 19-year-old boy doesn't have much to spare. It has to happen yesterday, if possible, because life is getting away from you. Add to that my peculiar learning difficulty, peculiar because I have a quick mind when it comes to language, and disaster was inevitable. I really should have been a thespian. When a joint didn't fit properly, I bashed it with my hammer and stuck in more glue, plus I never finished one job before moving on to the next.

Wharram's design and construction philosophy became more sophisticated later, partially a natural evolution when Hanneke Boon brought her artistic talents to the team, and partially a response to people like myself, who misunderstood his earlier messaging that the boats could be built by anyone, anywhere. We took it to heart and slapped together abominations.

It must be said that James bore some responsibility for the cavalier approach of builders like me. I remember writing to him during the construction, worried because one of my hulls, still in the framing stage, appeared to have a twist in it. He told me not to worry, that it would all pull straight when I fitted the plywood skin. 'I was once taken 400 miles by a worried Oro builder,' he wrote, 'to have a look at the framework of his hulls because they were not fair. I just gave them one almighty kick and everything straightened out.'

Anyway, John Guzzwell had written about his imperfect workmanship, bent nails and splices that were a bit rough, etc, in *Trekka Round the World*. I later discovered that he was a modest perfectionist. *Trekka* had been crafted like a grand piano. *Wicked things, stories.*

Unfortunately, the twists did not pull out of my hulls when I planked them, and the despondency this provoked probably accounted for some of the debacle that followed. Maybe I didn't kick them hard enough before putting the plywood sheathing on. Wharram had said you didn't need a strong-back, just set the frames up on the ground, but he probably expected more attention to detail, a string-line or something. I felt ashamed, looking at the twisted hulls, and covered them up when people came around.

Instead of taking stock, however, I seemed to lose whatever tenuous faith I had in myself, the dream slid into nightmare, and the workmanship deteriorated further. One night, one of the hulls blew over onto its side in a gale, and the small cabin prised itself loose from the deck. I shoved more glue into the gap and nailed it back down.

The next development is hard to write about but difficult to avoid. At some point after I started visiting the main public library in my mid-teens, I made friends with one of the female librarians, Patricia Felix, a buxom woman with long, blonde hair and Spanish heritage. I am not sure of Patricia's age, but she was several years older than me.

She used to dig archived books out of the stack room for me, and do research in various journals when I asked her to. She took me seriously, seemed to enjoy my company, and I became extremely fond of her. When I reluctantly went off in July 1970 to do 12 months national service in the army, first in Pretoria, and then in Bloemfontein, as far away from the sea as it is possible to get, we started exchanging letters. She had sensuous, flowing handwriting, and her words were filled with warmth. I was desperate for affection, and fell in love with her letters.

When I mentioned that I planned to sail away from South Africa after finishing my catamaran, Patricia replied, 'I am going to miss you.' I suggested she come along, without thinking of the consequences. She said she'd love to, but her parents, being devout Catholics, would not approve if we were not married. So, I proposed. I did it in good faith, but the physical reality was hopeless, and I caused her a lot of pain. In the long run, the situation became a catalyst that propelled her to some interesting adventures, and later a happy marriage, but I hurt her deeply and it does not excuse my behaviour.

To avoid the looming issue of sex, I buried myself in work. I was working 18-hour days on the boat now, having given up my job. Patricia was providing the money. Sometimes, despite watching every dollar, she dragged me out to dinner, to do the young couple thing. We usually went to the Point Yacht Club, where I could gaze out at the harbour through the plate-glass windows of the dining room. I loved the reflection of the city lights on the dark waters of the bay, and the ghostly outlines of the round-the-world yachts at the International Jetty. It was still my favourite place in the world, apart from the West Indies of course, Tahiti, New Zealand, or wherever. When I was down there among the voyaging yachts, the future seemed possible.

On weekends, there was a band playing and we danced. Patricia liked the slow numbers, when we could smooch around the floor. I tolerated this, though she once complained that I was as stiff as cardboard, and I rescued my honour by making a smutty joke. I preferred it when the music was hopping. Then I'd really jive, man, all arms and legs. I had good rhythm, would have made a fabulous professional dancer if fate had been kinder. I was proud of my dancing. Sometimes I had the feeling that if I just danced a little faster, a little wilder, a little longer, everything would turn out all right.

Once, I was dancing like this and people began to clap, so I danced a little wilder, a little faster. They clapped more. I was dancing like a helicopter. Then, as I flung my head back, exuberant, reckless, I saw the old man behind me. A well-known character in the club, always dressed in immaculate white shirts and cream slacks, with the best leather shoes, he was dancing with a girl who looked younger than me. The music was really hopping, but the silver-haired old bastard swung slowly across the floor with a knowing stateliness, smirking. I froze. It was like having my balloon popped. Then I shrugged and continued dancing, but the joy was gone. I felt like running over and head-butting him. *Time to go home.*

Pleas for me to delay launching until after the wedding ceremony were ignored. It was mid-December already and every day counted, as I planned to leave in January. Harry and Adrienne had sailed for Cape Town and the West Indies recently aboard *Kelasa,* and I was keen to catch up.

*Otaha* was launched a few days before the wedding. The name *Otaha* was an ancient Polynesian honorific, given to *Sandefjord* during its stopover in Tahiti. It means, *Great White Bird*, according to the *Sandefjord* movie. *Otaha* was white, with royal blue bulwarks. The boat was still half-painted, though, with some areas still just primed, and no antifouling paint on the bottom. As soon as we tied up, I carried on painting.

*Otaha motoring away from the dock after the launch. I am on the tiller and Patricia is walking across the deck. Photo: L.J.A. Cox (my father).*

A couple of hours before the wedding ceremony, I was still painting *Otaha's* bulwarks. My brother, Malcolm, finally dragged me off to get dressed. The ceremony was a small affair, mostly immediate family. My only guests were Lawrie and Lorna Williams. The pastor, an amusing Australian who'd once been a fairground boxer, and had the nose to prove it, held my paint-splattered hand and said admiringly, 'That's a nice shade of blue'. He was wearing a maroon shirt and paisley tie, and confided to me that he had his bathing trunks on under his suit, as he was heading off for a quick surf later, after he'd conducted a funeral. I'd been introduced to this splendid fellow by Lawrie and Lorna Williams. After the nuptials, when Patricia and I turned to face the congregation, my mother told me to tuck in my shirt.

Then it was back to *Otaha*, to hurry up and chase *Kelasa*. I still had a clear memory of the day Harry and Adrienne left. I had decided to run to the harbour entrance, about eight kilometres away, to wave farewell. As soon as *Kelasa* steamed away from the International Jetty in a cloud of petrol fumes, after Harry had spent a considerable time swinging Vivienne's crank handle like a maniac, sweating and swearing, to get the old engine started, I ran like a dervish through the car park and out along the harbour wall towards the T-Jetty. As I dodged trucks, cranes and crates of cargo, I caught glimpses of *Kelasa* in the gaps between moored ships. Sometimes I seemed to be gaining, sometimes losing. For a long time, I was losing, and it seemed likely that *Kelasa* would get there before me. I ran harder, lungs bursting. *I was gaining.*

*Harry conducting last-minute checks the day before Kelasa sailed from Durban. Note the new, flax mainsail.*

Then I was there, ahead of Harry and Adrienne. I felt exhilarated, even though I was so breathless I could not have spoken a word for several minutes. I enjoyed coming out onto the harbour breakwaters. It was almost like being on a boat, with the water sucking and sighing all around, and yachts sometimes passing by.

There was great drama as *Kelasa* motored out through the breakwaters into a brisk headwind, pitching heavily into the short, steep waves. Harry stood resolutely at the tiller, staring ahead, oblivious to the world, while Adrienne, sitting on the coachhouse, gave me a wave. Before they got clear, the engine stopped and *Kelasa* fell off towards the northern breakwater, on which I was standing. It later transpired there was water in the petrol they had recently bought. Harry blamed the supplier, but he'd left the clear plastic fuel drums on deck for several days before departure, which most likely caused condensation to form in the containers.

Adrienne took the tiller and Harry sprang for the halyards. He was shouting for Adrienne to turn back as he raised the staysail, but she insisted he hoist the mainsail and beat out. As Harry had said more than once, you don't argue with Adrienne, so he did as he was told, got the large, heavy, gaff mainsail up in astonishingly quick time, and then unfurled the jib, which was on a Wickham-Martin furling drum. *Kelasa* gathered way and cleared the outer end of the breakwater, on which I was standing, by about three boat-lengths. I was cursing almost as much as Harry, because I had not brought my camera, wanting to run as fast as I could.

*Kelasa* had been in Durban for my entire adult life at that stage, all two years of it, and Harry and Adrienne had become significant players in my world. I could barely imagine them no longer being around, and was desperately sad to see them go. Saying goodbye to Keith Kibler, Croix Grut, and Curtis de Camp had been hard enough, but this time the sense of loss was almost unbearable. Theoretically, I was soon to follow in *Kelasa's* wake, but my emotional turmoil suggested my subconscious knew better.

# Chapter Seven

# ...a Shadow Falls

Another reason to sail from Durban as quickly as possible was my desire to rendezvous with Keith Kibler, who'd just sent me a letter from Panama. It had been more than two years since I'd received news of *Korsar*. Frustrated, I'd sent off letters to several locations, hoping that one of them might get through. In *Cape Horn: The Logical Route*, Moitessier had described doing just that, when he was trying to find his friend, Pierre Deshumeurs, in France. And it worked, just as it had for Moitessier.

Keith was anchored off the island of Tobago, at the Pacific side of the Panama Canal. Felicia had gone back to live in the USA, and he had a new girlfriend now, Gloria. They were considering whether to head for Patagonia on *Korsar*, or back to the Caribbean, with the idea of trading in the old boat for a fibreglass vessel. *Korsar* was getting tired, he said. He gave me the address of his brother, and I replied immediately, but never heard from him again.

I had also been getting letters from Croix over the previous two years. He and Merle had sailed up to England, then on to Tahiti via the Caribbean and Panama. In a letter from Papeete, Croix mentioned that they had seen Moitessier and *Joshua* there. Moitessier, Croix said, had taken up with another woman, and was smoking pot with hippies. Perhaps remembering *Cape Horn: The Logical Route*, and that book's sweet evocation of Moitessier's love for his wife, Francoise, Croix wrote: *You do not find the answers to life by spiralling downwards.* Later, some of Moitessier's friends told me that *The Logical Route* was the 'honeymoon book', and did not reflect the 'true' Bernard.

*Iorana* had just arrived back home in New Zealand, and in Croix's latest letter, he assured me of a warm welcome when I arrived. Their circumnavigation had been flawless, without a hitch. They'd even passed through the Tuamotu Islands in French Polynesia, the so-called Dangerous Archipelago, with its low-lying coral reefs and infrequent lighthouses, during the night. They hove-to for a while after the weather broke off Takaroa Atoll, before star sights gave them a good fix at 2200, when the sky cleared briefly. A good cocked hat (intersecting position lines from at least three stars) established their position south of Arutua Atoll, allowing them to ease sheets for Tahiti.

*Iorana moored in Robinson's Cove, Moorea, Society Islands, July 1971, with Croix contemplating the scene. Photo: Merle Grut.*

I loved getting those letters, with their colourful stamps and the news of *Iorana's* travels. I still have them, and Keith's, tucked safely away. Sometimes I re-read them, and remember the warmth of our friendship. Keith and Croix have both sailed for Valhalla now.

After *Otaha* was launched up at the Bayhead, near all those old wrecks I used to dream about, I had motored across the bay in triumph to the International Jetty, although triumph is perhaps not the right term. Our old, underpowered Seagull outboard motor, spluttering and smoking, barely stemmed wind and tide. At one stage we had to jump overboard and push *Otaha* off a sandbank.

At the jetty, I tied up next to a 7.09m, 1200kg Spanish yacht, *Mistral*, a flimsy fibreglass day-sailer being sailed around the world by 28-year-old Julio Villar, who, I thought, had the longest, curly brown hair I had ever seen on a man, apart from magazine photos, although it does not look that outrageous to me today.

*Julio Villar on the deck of Mistral in Durban, December 1971.*

He'd sailed from Spain at the age of 23 after a horrific fall terminated his impressive mountaineering career. He had a scar that went from his right collar bone to below his left breast, then around the back and up to his shoulder blades, but he looked muscular and fit. He was the most carefree guy I'd ever met, spending his days playing harmonica and flirting with women. He had girlfriends in every port.

Sometimes he made very slow passages. He took so long to sail down from Mauritius to Durban, 40 days from memory, to make good 1700nm, that his friends declared him overdue. He sailed blithely in several days later. When questioned, he just shrugged. 'The wind was too strong,' he said, 'so I just took the sails down and drifted. I cannot bash against heavy seas.'

That was the secret of his success. To quote an old Chinese saying: *If you understand your weakness, it can become a great strength.* On passage to New Zealand, he came up against strong westerly winds when closing the coast. After drifting for 10 days, unable to make headway against wind and sea, he gave up and sailed back to Fiji, reprovisioned, and set out again.

*Mistral in Cape Town, February 1972. Note the double forestays for his twin jibs, which were tied on either side of the pulpit railings. I never saw the boat with sailcovers fitted, if Julio had any.*

He never did a day's work on the boat. *Mistral's* fibreglass gel-coat was cracked, faded, and streaked with rust-stains, but he flatly refused to paint it. 'Once you paint a boat you always have to,' he declared, with indisputable logic.

*Mistral* had a wind-driven self-steering gear designed by Bernard Moitessier. In the years after the Golden Globe Race, Moitessier had designed several of these gears for penniless vagabonds who passed through Tahiti. Before Moitessier designed *Mistral*'s self-steering gear, Julio said, laughing, he had just gone to bed at night and let the boat sail where it wanted to.

He sold his outboard motor in Durban to give him funds for the last leg home, and set off for Cape Town with an aerial photograph, torn from a *Life* magazine, as his only chart of that harbour. Before he left, he careened the boat against the rickety jetty on the other side of the Point Yacht Club, and discovered two bulges below the waterline, egg-sized hernias, where the twin mast-compression posts were driving through the hull. I seem to remember that there might have been a missing bulkhead involved, removed to gain more legroom below. Undismayed, he put a third post between the existing two.

'They got me three-quarters of the way around the world,' he said with a grin, 'so I think this will get me back.' And it did. Over the course of a four-and-a-half-year circumnavigation, he sailed 38,000nm and crossed the Atlantic Ocean three times. When last heard of a few years ago, he still owned *Mistral*, now lovingly restored and smartly painted. He wrote a book in Spanish that developed something of a cult following, *Eh, petrel!*, but it has never been translated to English. I have a copy, plus a Spanish-English dictionary, and have read a few pages, but it remains a retirement project.

I remember with amusement the day I asked Julio how to cook rice. I meant that exotic Spanish dish, Paella, but he misunderstood. "Ah, Graham,' he sighed, 'you have so much to learn.' And that was true enough.

Shortly after we rafted *Otaha* up to *Mistral*, the furious Assistant Port Captain came to visit. 'Who gave you permission to launch this boat?' he demanded. In Durban, you had to get permission to launch a local boat into the harbour, as there was a waiting list. I had not asked because I considered it my right. After all, the voyaging yachts just sailed in from wherever. I was a voyager too, as I saw it, I just hadn't gone anywhere yet. The Assistant Port Captain, who didn't see it that way, walked around the decks, looking scornful. He kicked the cabin side.

'Hey,' I shouted, perhaps thinking of the time the cabin had come adrift from the deck when one hull fell over before I'd joined them together.

He looked up sharply. 'I will never give you permission to take this boat out to sea. It's a floating death trap. It wouldn't last one day out there'.

'What would you know?' I sneered. I wanted to push him off *Otaha's* deck, but knew that would be foolhardy.

'Well, I know this,' he said, 'Both of your parents,' he nodded at Patricia and me, 'have spoken to the Port Captain about you, and they have pleaded with us not to let you go.'

'You can't stop me,' I sneered. 'I'll sail out in the middle of the night.'

He smiled, as if he was listening to an imbecile. 'Nobody could possibly sneak in or out of this port without being apprehended, young man. If they could, this place would be crawling with communists.' He shook his head. 'What's more, your parents are prepared to commit you for psychiatric treatment if you persist.'

I looked wildly to Patricia for support, but she was looking at the deck. I had a sudden moment of clarity. I don't think she ever intended to sail on *Otaha*. It was all over. I turned away, went inside one of the hulls and lay on the narrow bunk. In a parallel universe, the other Graham knew that the boat was a death trap, anyway.

One day I went down to the breakwater and watched *Mistral* sail away in company with three other yachts, *Ophelie*, *Moala*, and *Ghost Rider*. There was a fresh northeaster blowing straight in through the breakwaters, kicking up a short, steep sea over the bar, as the four boats tacked out of the harbour.

Sometimes Julio came so close to the rocks I could almost reach out and touch him. He was his usual, insouciant self, laughing and cracking jokes. The skippers of the other yachts looked more subdued, as well they might, setting off on this notorious passage. I sat on the rocks, surrounded by crashing waves, the tang of salt-laden air, and raucous gulls, watching until Julio and the others merged with the horizon, until I could no longer differentiate them from the specks dancing in my vision. I could not comprehend the shadow that had fallen across my dream.

*Mistral sailing out of Cape Town yacht basin. Peter Strong's Wayfarer is up on the dock in the distance. Note Bernard Moitessier's self-steering gear.*

I was so disturbed by seeing them leave without me that I followed them overland to Cape Town and stayed until they set sail, in company once more, for the South Atlantic trade winds, Saint Helena Island and beyond, taking a piece of my heart with them.

All four boats made trouble-free passages to Cape Town. In fact, during the years that I haunted Durban's International Jetty, not a single visiting yacht got into difficulties sailing around the Cape. Combined with the laconic attitude of their skippers, and my wide reading of the sailing classics, I formed the opinion that tradewind ocean voyaging was safer than living ashore, and that you could do it in just about anything that floated. Julio's aerial photograph of Cape Town had served him perfectly well to guide him into port, and the lack of an outboard motor did not seem to trouble him in the slightest.

*Ophelie following Mistral out of Cape Town yacht basin.*

*Ophelie* was an 11.58m ketch from Meta Boatyard, the French builder of *Joshua*. The 29-year-old skipper, Yves Jonville, was just as enigmatic as Moitessier was said to be. He didn't like it when people compared *Ophelie* to *Joshua*, but the boat's details were very similar, right down to the colour scheme, with bulwarks and welded deck fittings painted black, contrasting with the white decks; except that the hull was yellow and had a small transom. The Jonvilles were on the last leg of a circumnavigation, and had first met *Mistral* in 1968, in the Atlantic Ocean's Canary Islands. They had crossed three oceans since then, meeting up in various ports, and had developed a tradition of leaving port together. In the Indian Ocean, they'd been joined in this tradition by *Moala* and *Ghost Rider*.

*Graham Tait, being cautious by nature, decided to motor Moala clear of the moorings in the yacht basin before hoisting sail.*

*Moala* was a 34', carvel-planked ketch with raised topsides, a typical Maurice Griffiths design, registered in Melbourne, Australia, and sailed solo by a quiet, cautious Australian named Graham Tait, who spent six years building the boat in his backyard. He professed to getting nervous before setting off on a passage, but said that once underway he was OK. He was heading to England, where he thought he might sell *Moala*, as he was engaged to be married there, but a couple of years later I read in *Pacific Islands Monthly* that he had arrived in Rarotonga in the Cook Islands, on his way home. *Moala* had sprung a bad leak after leaving Tahiti, forcing him to pump almost non-stop.

*Ghost Rider* was a 24' flush-decked cutter from Sydney, Australia, custom-built from an Endeavour 24 fibreglass hull. The boat was quite strong, though putting a flush deck on it resulted in a tiny interior.

*Ghost Rider, with Wendy on deck, sailing out of the Cape Town yacht basin.*

*Ghost Rider's* skipper, Steve Dolby, was a 26-year-old Englishman who was slowly going blind after a surfing accident, according to the Durban newspapers. He drove *Ghost Rider* hard, and had made fantastic times in the Indian Ocean trade winds, surfing regularly, until the boat broached on the face of a wave between Cocos Keeling Atoll and Mauritius, sticking the spinnaker pole into the sea, and breaking the mast.

Steve arrived in Durban solo but left with a young woman called Wendy. She left *Ghost Rider* in the Caribbean, and Steve sailed on alone to New York. A few years later, he customised a 26' fibreglass yacht in England, though he did not flush-deck it this time, named it *Ghost Rider II,* and sailed solo to Australia, thus completing a personal circumnavigation.

Another yacht in Cape Town that I'd first met in Durban was the 32' carvel-planked American cutter, *Island Childe*, skipped by a retired Los Angeles policeman, Gene Davidson.

*Gene Davidson in the cockpit of Island Childe in Durban.*

I found it hard to reconcile genial Gene, who often had young, long-haired, scruffy crew aboard, with the hard-nosed reputation of the Los Angeles law enforcement department. I remember Gene saying you had to keep your hand on your gun when approaching suspicious vehicles on the Los Angeles freeways, because too many officers had been ambushed. He was thoroughly enjoying his retirement, and 'seeing the other side of life', as he put it.

After Gene returned to California, he wrote me a letter saying that Julio had spent a month on Saint Helena Island, where he had 15 girlfriends, because many of the young men worked away on Ascension Island to the north. Julio then turned up in Recife, Brazil, while *Island Childe* was there, before heading up into the North Atlantic and on to Spain, to complete his 38,000nm circumnavigation. Julio's route avoided bashing to windward against the NE trade winds on the direct route via the Azores, another example of his strategic approach to voyaging aboard his small, fragile yacht.

Gene then sailed up past the Amazon River, where he ran aground on a sandbank, to the Caribbean. *Island Childe* was knocked down in a storm on the way to Panama, without significant damage, and made it safely back to San Francisco, still with his loyal crew, Bill, a Canadian who'd joined in Durban. In the conclusion to his letter, Gene said, 'The land looks good to me.'

*Clipper motoring out through the Durban breakwaters en route to Cape Town, at the start of Clive and Margaret's voyage to Australia via the Caribbean, Panama, the South Pacific and New Zealand. Photo: courtesy Margaret Worthington.*

Clive Rouse also sailed away that summer. While I was in the army, he'd cruised *Clipper* to the Seychelles Archipelago and back via the Mozambique Channel and Madagascar, with a young English girl as crew. He spent 1971 refining *Clipper*, during which time he met Margaret Worthington, a local artist a couple of years older than me, and she joined him for the voyage.

I made my usual pilgrimage to the harbour entrance and watched *Clipper* sail away. It was a particularly emotional experience, as *Clipper* was the first yacht I'd ever sailed on, and for a while there, I thought Clive and I were going to embark on a great adventure together.

Clive and Margaret were in Cape Town when I visited, after a rough trip around the Cape of Good Hope, in which *Clipper* had been knocked down and broke its tiller. The broken tiller was still lashed to the rudder head when I visited them. Clive had been attempting to round up to lie ahull when a breaking wave caught them, and for a while both he and Margaret were underwater. Margaret was undaunted by the experience, and was eagerly looking forward to further adventures.

*Clipper locks through the Panama Canal in 1972. Margaret stands by the mast while Clive is on the tiller. Photo: courtesy Margaret Worthington.*

It was another 11 years before I saw *Clipper* again, and even longer before I caught up with Clive and Margaret. They crossed the South Atlantic to the Caribbean, transited the Panama Canal early in 1973, and went on to New Zealand. In 1975, they sailed on to Australia via the South Pacific, settling in Gladstone.

There was a famous old schooner in the Cape Town yacht basin, the *Dwyn Wen*, built by Phillip and Sons in the UK in 1906, 106' on deck and displacing 150 tons. Once a gentleman's yacht, it was now working in the charter industry, and en route from the Caribbean to the Seychelles. It had obviously seen better days, but was still an awesome sight, filled with grace and power.

I was talking to a member of the ship's company one day, Phil Wade, who was either the mate or skipper, I forget, and he surprised me by saying he'd sailed from Cape Town to Rio de Janeiro aboard *Ninetails* with Tom Corkill. According to Phil, it was an alarming experience, as the cat creaked and flexed all the way, making him fear it might break up. Perhaps Phil wasn't used to catamarans, or Tom's devil-may-care attitude saw him through with this boat, but his subsequent history with *Ninetails* suggests the boat was more seaworthy than Phil assessed.

*The 106' schooner, Dwyn Wen, called at Cape Town while I was there, en route from the Caribbean to the Seychelles.*

A few weeks after we returned to Durban, Patricia and I went to help an older friend of mine, Ron Glover, work on the ferro-cement yacht he was building. I soon went back to the yacht basin alone. Patricia and I were divorced within six months. She stayed on with Ron, completed the yacht, named *Pambili*, and sailed to South America and back with him. Later, she met another man and they married in 1981.

I sold *Otaha* for SAR 700, about one quarter of what Patricia and I had spent, and moved aboard an abandoned yacht, *British Viking*, at the International Jetty, sharing it with a bunch of dope-smoking American hippies. *Spiralling downwards*, Croix might have called it, but they were there for me, and I appreciated their fellowship. My only other option was to return to my parents' house, and I had no intention of doing that. I should have gone back to work, and in those days it was there for the taking. It would have given me options, but I felt deflated, with no direction home.

I should have given half the money to Patricia, but justified my needs as being greater than hers, and she was gracious enough not to press a claim. I did give her the two gimballed, brass lamps we'd received as a wedding present, even though I coveted them.

My father came to see me occasionally. He stood awkwardly on the dock and asked how things were, but didn't offer any advice, for which I was grateful.

*Otaha on the foreshore in Dead End Creek. Ron Glover's son, Toby, stands on deck.*

At about this time, I was befriended by Australian circumnavigator, John Murray, on his cold-moulded plywood trimaran, *Unbound*, a 36' version of Tom Corkill's *Clipper*. He'd named his boat *Unbound* because he was not going anywhere in particular, but I teased him, saying I hoped the three hulls would not become unbound and sail off in different directions. Some multihull enthusiasts might have taken offence, for this was an era when multihull seaworthiness was hotly contested, but John just laughed. He was a large-limbed, ginger-haired, freckle-faced larrikin, who loved cocking a snoot at pompous bastards. He encouraged me to take my boat up the coast and launch it through the surf, but he hadn't had a close look at it.

The sane Graham had almost merged at this stage with the crazy one. I was easing back into my skin, although it wasn't to be the last time I had a period of temporary insanity, when stress seemed to divide the self. It was fun just lounging around the yacht basin. The hippies introduced me to marijuana, and I'd sit for hours in the cabin of the derelict boat, stoned off my face, listening to them tell jokes, play guitar, and mouth off about fascists. I discovered that I disliked marijuana, as it made me paranoid, and I soon stopped smoking it.

John Murray took seven years to circumnavigate out of Sydney, between 1969 and 1975. *Unbound* had a 10hp diesel engine but it seldom worked, and he sailed engineless for much of his voyage. There was also no radio transmitter, refrigerator, or liferaft. Basic boat.

*Unbound alongside the International Jetty. Note the repair to the starboard side. Unbound was built in the open and John was always chasing rot.*

After rounding the Cape of Good Hope and cruising extensively in the Caribbean, John sailed *Unbound* up to New York with his South African partner, Florence, who he met in Durban. They explored the Inland Waterway and the Bahamas, before eventually heading for Panama and the South Pacific Ocean. I remember John's story of sailing between the Galapagos and Marquesas Islands, in the heart of the SE trade winds, where at one point they did not touch the tiller, sheets or sails for 10 days straight.

*Unbound* had been built in the open, under a tarp some of the time, and John paid the price by periodically cutting out and repairing patches of rot. He also ran aground on a reef in the remote Louisiade Archipelago of Papua New Guinea during the first year of his voyage, tearing the bottom out of the main hull. With the assistance of local villagers, he floated *Unbound* across the lagoon, jacked it up out of the water onto a bamboo cradle, and spent a year rebuilding the hull with hand tools. All of his materials had to come from faraway Port Moresby, and often took weeks to arrive. He also took the opportunity to install a deep centreboard, which transformed the boat's performance. Years later, when I sailed aboard *Unbound*, it was like handling a large dinghy.

Lucy and Jeff Montigue, who circumnavigated on their Hedley Nicol trimaran, *Kunda Wonga*, very similar to *Unbound,* were also in Durban in 1971. They returned to Australia before John, becoming the first Australian multihull to sail around the world.

*Jeff and Lucy Montague's trimaran, Kunda Wonga, the first Australian multihull to circumnavigate, anchored in front of the Point Yacht Club.*

I remember the day that Jeff invited me aboard *Kunda Wonga,* moored fore and aft just off the Point Yacht Club. The water was shallow there at low tide, so he had the space all to himself, one of the advantages of multihulls. As I stepped across the side deck into the cockpit, I was startled by a loud crack beneath my feet. I stopped in mid-stride, shock and fear coursing through me. I was always breaking things. Clumsy oaf.

Jeff grinned, looking amused. 'Don't worry about that,' he said. 'It always cracks.'

After safely rounding the Cape of Good Hope and making landfall on St Helena Island, Jeff and Lucy hardened up the sheets for a beat across the NE trade winds, bound for the Azores Islands and England. After a week of beating into the wind, the sound of cracking from the deck became so loud that they decided it would be foolish to press on. They eased sheets and sailed downwind to the Caribbean, where they rebuilt the decks. Then they proceeded with their voyage, eventually returning to Brisbane via Panama and the South Pacific.

One notable feature of *Unbound* was that it had a wind-operated self-steering gear, with a windvane driving a trim-tab on the trailing edge of the outboard rudder, though the design did not have a differential linkage and over-steered somewhat. Wind-driven self-steering gears have had a chequered history on multihulls, because these boats often sail very fast off the wind, surfing on wave crests, which causes the direction of the apparent wind to vary widely and confuse the windvane. Tom Corkill did not have one on *Clipper*, which almost certainly contributed to his disaster.

John had some success with his home-built unit by keeping the boat-speed down to seven knots on passages. On those rare occasions when he drove the boat beyond hull-speed, usually daysailing inshore, he would hand steer. He is one of the best seamen I have sailed with, and I always felt safe sailing with him on *Unbound*. He commanded the deck, and his friends still call him Commander Murray.

One morning when I stuck my head out of the hatch, I was delighted to see a 32' gaff-rigged ketch called *Walkabout* moored nearby. I had first come across a photo of this boat in Eric Hiscock's book, *Voyaging Under Sail*. At the time Hiscock photographed it, in 1954, the boat was in Durban, but nobody could tell me where it had gone. It was designed by John Alden, a sistership to W. A. Robinson's famous *Svaap*, and had been built in Freemantle, Western Australia, from a dense hardwood called Jarrah. It was said to be massively strong, very heavy, fast in strong winds, and wet.

Mike and Liz Saunders found the boat in a remote part of Mozambique, fitted it out with some difficulty, and were en route to England with their four young children, making for quite a crowd aboard. I saw the boat again in Cape Town some months later, where they had spent the winter, after arriving too late in the Cape to voyage onwards, especially after the youngest boy, Bruce, fell and broke his arm.

The kids were a lively bunch and seemed to enjoy an enviable freedom. A few years later, in 1976, I bought a copy of Mike Saunder's' book, *The Walkabouts*, which remains one of my treasured titles. It is full of infectious good spirits and humour, and Mike succeeds in bringing the characters, situations and places to life in a delightfully vibrant manner.

*Walkabout close reaching. Note the vang from the end of the mainsail gaff to the head of the mizzenmast, preventing the gaff from sagging off, and dramatically improving windward ability. Photo: Colin Harrington. Image scanned from Mike Saunders' book, The Walkabouts.*

Months went by, my hair grew down to my shoulders, and I had no idea of what to do next. One morning when I was rowing around the harbour in my little inflatable boat, which tried to fold up every time that I pulled on the oars, John called me over to *Unbound*.

'Come and have a Bloody Mary,' he said, even though it was not yet lunch time, let alone sun over the yardarm time. 'Listen, young cobber,' he added, once he'd lubricated my cognitive functions with Vodka and tomato juice, 'it's about time you decided what you're going to do with your life. You can't sit around getting stoned with a bunch of hippies forever. Why don't you go to Australia? There's lots of opportunity for keen young fellows like you there.'

'They wouldn't let my boat out of the harbour.'

'Well, I offered to help you launch it through the surf.'

I shrugged.

'How much money have you got?' he asked.

'About three hundred rands.'

'You can get a cheap ticket to Australia for less than that. One of my mates just took a ship back to Sydney from Cape Town, and it only cost him two hundred. Here, I'll give you the address of the shipping agent.'

*Young and fresh faced, about to leave Durban for a new life in Australia.*

So that's what I did. Having been told that Australian officials didn't like *long-hairs*, I traded in my beautiful hippy hair for a savage crew-cut, and bought myself a one-way ticket from Cape Town to Sydney. I arranged to spend the last evening at my parents' house, since my brother, Malcolm, offered to drive me to Cape Town.

It seems obvious now that my mother would cook a special farewell dinner, but it didn't occur to me then. When John Murray threw a surprise farewell party for me beneath the hulls of *Unbound*, which was out of the water in the car park of the Point

Yacht Club for a refit, I never gave my parents another thought. The party raged long into the night and was still going strong when I had to leave to catch the last bus home. As I departed, John raised his beer bottle in salute. 'Cop you later,' he shouted, with a cheeky smirk. It was almost six years before I saw him again.

It was past midnight when I got home, drunk and covered in mud, after falling in a ditch. The house was in darkness as I stumbled in and collapsed for the last time on my childhood bed. The next morning, hung-over and withdrawn, I quickly packed a suitcase, cramming in a few clothes and as many of my favourite sailing books as possible. My mother later posted the rest to me at considerable expense to herself. I didn't feel like eating breakfast, so went out onto the verandah to wait for Malcolm. Poor Alec, my little brother and lifetime buddy, got a brief nod. I never thought about what this departure might mean to him.

My father shook my hand, saying, 'Just promise me you'll write to your mother. Women are different to us. They give birth to you and find it hard to say goodbye.' I only realised a decade later that he was the one who was heartbroken. My mother was always self-contained. She gave me a brief hug. She'd never been big on hugs, except with my father. She still sat on his lap sometimes, after 25 years of marriage. *It's crazy the things you think of at times like this.*

As the car began to move, I glanced up through the passenger window, to where my parents stood on the back verandah with their arms around each other, looking down at me. *Perhaps it's just the unusual angle,* I thought, *but it looks like they are clinging to each other*. I gave them a little wave. Mindfulness was not one of my best attributes.

# Chapter Eight

# Chance is a Fine Thing

It's a good thing I did not know what lay ahead. The justly famous Sydney Harbour is beautiful, but the city has a harsh underbelly, and is no place for the down and out, as I quickly discovered. Perhaps if I had gone to one of the voyaging crossroads, such as Whangarei or Papeete, or even Brisbane, just up the coast, I may have found some continuity with the happy days spent on the International Jetty in Durban, surrounded by yachts from all over the world, thrilling to the extraordinary camaraderie I found there.

Cruising yachts have always visited Sydney, but they scatter around the vast harbour. In time, I'd find a marina that some favoured, where a smidgen of the old atmosphere was rekindled, but it took years for me to stop grieving for the magic of Durban's yacht mole.

Not that I had a choice of where to go. Britain, Canada, the USA, and my first choice, New Zealand, all rejected my applications for visas. They did not need unskilled workers, even if I was just a boy with who knows what potential, and Britain was positively hostile to anti-apartheid activists. Australia was the only country that advised me I was welcome, as long as I had a current South African passport. Having completed my national service, that shouldn't have been a problem, but the South African Government were still reluctant to issue me with one. They eventually gave me a non-renewable passport that was only valid for a period of three years, and restricted to Australasia.

I was advised on arrival in Australia that I could apply for citizenship after a period of three years, and that it would be granted provided I had a guarantee of employment and did not have a criminal record. As the smiling Immigration officer said to me, 'You don't look like you robbed the Bank of Monte Carlo, son.' Being young and fresh-faced has its advantages. I counted down the days.

It took 20 days for the *Patris*, an old, Greek passenger ship, to sail from Cape Town to Sydney, via Mauritius, Perth and Melbourne. There were several hundred Greek migrants on board and a handful of English-speaking adventurers. I'm not sure what category I belonged to. I stood for hours at the stern, watching the foam twist off the propellers and disappear into the vast ocean, trying to imagine my heroes who had sailed these waters, to see myself out there on my little sloop. I was in a daze, sure that I was going to be sent back to South Africa when the ship arrived in Australia, despite the advice I had been given. I never went up to the bows.

Mauritius was a bright, colourful, lush island, with jagged mountains, exotic wildflowers, passion, poverty and squalor, although I only saw Port Louis. I met a French sailor there, Henri Cordovero, on the 9.75m cold-moulded timber sloop, *Challenge*. He was an old friend of Bernard Moitessier's, having spent time with him in the Mediterranean, not long after *Joshua* was launched, and more recently in Tahiti. I gave him advice about Durban, and he gave me the address of a young French couple who lived in Sydney, and who had recently bought a yacht.

Sadly, *Challenge* capsized en route to Durban, with the loss of a young crewmember who'd joined ship in Port Louis. Henri eventually recovered, rebuilt the boat and sailed on, but I think the shock of such a thing never leaves you. He was a changed man the next time we crossed paths.

Perth, in comparison, seemed dull, predictable and flat, blasted by a relentless sun and scoured by the afternoon seabreeze. Drought-stricken for the previous seven years, the place had an air of isolated frustration. I had no desire to stay there, but appreciate my view was brief and biased. I was dreaming of tropical beaches and palm trees. I was also scared, with the first intimations of what I'd done, realising I was utterly alone, almost penniless, in a strange country.

Melbourne was dark and gloomy, with a cold southerly wind blowing directly off the Antarctic icecap. By the time I'd walked the length of St Kilda Pier, my hands were turning blue. When I asked the guy selling fish and chips in a stall at the end of the pier if it was always this cold, he just said, 'Yup.' He was probably joking, but at the time, I thought, 'Bugger this place.'

It was there that I saw my first *Hare Krishnas.* They have largely disappeared from Australian streets now. The familiar sound of jingling bells, drumbeats and chanting, was followed by a herd of them sweeping around a city corner, looking quite distinctive, with their robes, faces daubed with white clay, and shaved heads. Durban was full of Hindus, but it was the first time I had seen white ones. Astonished, I watched the troupe as it danced and sang its jubilant way through the midday crowds.

And suddenly it struck me. There were no black people in sight. I'd heard of Australian Aborigines, of course, but every picture I'd seen showed them in remote areas and deserts, almost naked, carrying spears and boomerangs. I assumed that was where they lived, and it was some time before I realised the situation was more complex. As Julio had so recently said, I had a lot to learn.

I will never forget my first glimpse of Sydney Harbour. It was spectacularly beautiful, bay after bay of burnished blue, framed by orange and grey sandstone cliffs, fringed with silky beaches and natural bushland, with thousands of yachts swinging to moorings, or fluttering around the tableau. Arching over it all, or so it seemed, was *the coat-hanger*, as the locals call Sydney Harbour Bridge, and the sensuous, shell-like curves of the as-yet-incomplete Opera House.

To my astonishment, I passed through Customs and Immigration without question, and found myself standing on the sidewalk wondering what to do next. I had A$45 in my pocket and no idea of how to present myself or where to look for work. If I had known more about the place, or been more worldly, I think I'd have secured factory or warehouse work without difficulty, because there was almost full employment at that time. Workplaces hung shingles outside their premises, listing positions available. All this was to change soon, when the 1974 oil crisis precipitated Australia's first economic recession since the boom that followed World War Two.

Instead, I became stuck in the inner city with rapidly dwindling finances. I ended up living on the streets, hanging out with a 16-year-old kid called Brett, who'd asked me for a dollar one evening and was shocked to discover he had more money than I did. I never got to know much about him, apart from the fact that he'd run away from an abusive step-father in a country town. He didn't trust people enough to open up emotionally. I was guarded, too, afraid that Brett might reject me if I revealed too much about myself. At night, we sometimes slept in a squat, though that was not always possible.

Most of the time I was hungry, though Brett showed me how to get free food from various sources, if you weren't fussy, but you couldn't just eat when you felt like it. Very rarely, as a special treat, I bought a plain hamburger. In those days they cost 25 cents, which seems little enough, but I couldn't afford it. My pitiful cash reserve was shrinking, and panic hovered just below consciousness. Unlike Brett, I was unwilling to beg, or do whatever else he did, to hustle a dollar. Despite Brett's anger and occasional despair, I sometimes envied him. He *belonged* on those streets, in a way I felt I never would.

I remember receiving letters from my mother and Croix Grut, care of the General Post Office in Martin Place, a grandiose building that reminded me of Durban City Hall. I read them on a bench outside the building, weeping, while workers on lunch break averted their gaze. It did not help that I was pretty unkempt by this stage. I felt lost, but something in me refused to ask for help from my family. And I was too embarrassed to open up to Croix. Whatever happened, there was no way I was going back to South Africa.

Brett often went off alone, to do some mysterious thing he would not talk about, and I'd wander around Kings Cross, which had once been a colourful, bohemian locale, with funky cafes, folk music clubs and art galleries, but was now just a sleazy dive full of organised crime, gamblers and prostitutes. I felt lost, as if a limb had been torn from my being. I realise today that I had suffered a form of depression throughout my childhood, but it had been disguised by hope. Now it bared its teeth.

One day while I was wandering aimlessly, I came to a sign, with a finger pointing down a hill, that said *Rushcutters Bay*. That was where Adrienne met Harry on *Kelasa,* and where *Sandefjord* moored during its circumnavigation, and reading the sign made my pulse quicken. After walking down the hill, I came to a large park, across which I could see masts in the distance. I hurried across the park and stood at a low seawall, which curved around the bay, looking at a slew of yachts, some swinging to moorings, some tied up at a long timber jetty that looked a bit like the International Jetty in Durban. I was skipping with excitement.

I hurried around the seawall to take a closer look, but was disappointed to find signs on a two-storey, red-brick building that proclaimed: *Cruising Yacht Club of Australia, Members Only, Trespassers will be prosecuted*. Bugger. I craned my neck through the gates but still couldn't see anything. There was a large slipway next to the club, with yachts hauled out, but a high fence blocked my view of that as well.

Later, I'd learn to disregard these signs and just waltz in past the slipway, as many people did, though it was never without a sense of trespass. As long as you didn't go into the clubhouse, nobody would challenge you. It was a more trusting society in those days, and security gates with electronic fobs had yet to be invented. On this day, however, I merely walked disconsolately along, hoping to find a better vantage point.

*The old timber jetty at the Cruising Yacht Club of Australia (CYCA), Rushcutters Bay.*

I'd just passed one gate when I heard a soft, diffident voice behind me that sounded vaguely familiar. 'Excuse me,' a male voice said, 'you're not looking for me, are you?'

I swung around, to find a short, craggy-faced man peering hopefully at me. It was David Lewis, whom I'd last seen in Durban in 1966, on his catamaran, *Rehu Moana*. I'd since read all of his books. David was waiting for someone called Peter Fry, an ABC journalist he'd met in Papua New Guinea several years earlier, but couldn't remember what he looked like. In 1966 I had only been 14, so perhaps my face *was* vaguely familiar. I'm sure I looked like I was in search of *something*.

It was also typical of David that he either did not notice, or did not care about my dirty, dishevelled appearance. He was an unconventional man in the purest sense of the word, equally at home with paupers and princes. He had just purchased a 32' steel sloop, that he'd renamed *Ice Bird*, on which he hoped to circumnavigate Antarctica singlehanded that summer. Nobody had attempted such a voyage before.

We stood talking for a while. There was no sign of Peter Fry (who arrived later), so David invited me to come and have a look at the boat. In through the unfriendly gates we went, down the old timber dock, past dozens of dazzling yachts. I recognised the wharf from the *Sandefjord* movie. Because of this perceptive filter, that wharf was always an exciting place to me, even though, socially, it would remain alien territory.

*Ice Bird* looked small for David's proposed voyage, and was entirely devoid of special gear. With its dark blue hull, pale blue decks and basic sloop rig, it looked like any other yacht used for week-end jaunts around the harbour. But if the famous Dr David Lewis thought it was suitable, it must be. It was a pretty boat, designed by Dick Taylor, an eminent structural engineer with a good eye for proportion, even if he was an amateur yacht designer.

*Refitting Ice Bird in Rushcutters Bay, early October, 1972. We have already bolted steel plates over the windows and welded others across the companionway, fitted a dodger over the main hatch, and mounted the undersized Hasler-Gibb servo-pendulum self-steering gear. Photo: David Lewis archives, courtesy Barry Lewis.*

Today, one could criticise the boat's narrow waterline beam and slack bilges, which combined to make *Ice Bird* tender, plus the inefficient rudder that was too far inboard, leading to steering issues downwind, but when the boat was built in 1962, those features were common in cruiser/racer yachts, especially those influenced by the English school.

When we climbed on board, David offered me an orange. Perhaps it was the way I tore it apart with my teeth that made him offer me another, then a slice of bread and cheese, then half a loaf. I wolfed it all down, as it was the first food I'd eaten that day.

David watched with interest, then said, as diffident as ever, 'If you've got some time to spare, I could really do with a hand to fit the boat out. I've got to leave in late October to catch the season, and there's an awful lot to do. But I can't afford to pay much.'

It was enough, nonetheless, for me to secure a furnished room in Surry Hills for A$10 a week, albeit a small, smelly and leaky one (Surry Hills was little more than a slum in those days), and to finally retrieve my suitcase from storage, with its precious cargo of books. It had been held by the People's Palace in George Street, where I'd spent my first couple of nights ashore. There were insufficient funds for electricity in my new digs, so I made do with candles. That meant the stove and fridge didn't work, but luckily David fed me.

I tried not to think about the future, happy enough for the time being to have one foot on the ladder, and to be back around yachts again. My only regret was that I could not persuade Brett to join me. I thought he'd be delighted to move into the flat but he refused. He'd been rejected all his short life and seemed to think it inevitable that I'd abandon him too, that I was already moving on. I think he felt safer in the world he knew.

'But it's fun hanging around the yachts,' I pleaded.

'I hate the sea,' he said. 'It makes things rust.'

By the time David sailed, Brett had disappeared, and I never found out what happened to him. After that, I stayed well clear of Kings Cross and Darlinghurst Road. It still makes me sad to go there, and I have no love for the inner-city suburbs of Sydney.

It seemed unbelievable that David could be serious about leaving so soon, given how much work there was to do, but he explained that the Antarctic summer was brief, and he needed to use it to its fullest. I had an enormous amount of fun, but several other people, including David's literary agent, Tim Curnow, with more responsibility perhaps, found the situation less entertaining.

It is easy to claim, as some have, that there was a breathtaking recklessness to David's preparations. And yet, he'd had an enormous amount of experience with sailing expeditions by then. The things he chose to do, within the temporal and financial limitations of the project, such as bolting steel plates over the portholes, and welding plates across the lower half of the companionway, ultimately ensured his survival, to which one could add his indomitable will, plus his ability to improvise and think strategically. Therein lay his ge nius.

He did admit to me later that he'd underestimated the Southern Ocean. While crossing the North Atlantic Ocean in high latitudes aboard *Cardinal Vertue,* returning from the

1960 OSTAR, he'd experienced a number of severe autumn gales, and coped well. The North Atlantic in autumn has a dreadful reputation. Only mad dogs and Englishmen venture out there in small yachts at that time of year. David had been confident he could manage aboard *Ice Bird*, but found the fetch and ferocity of the Southern Ocean beyond his imagination.

Moored near *Ice Bird* was a 30' catamaran called *Annaliese*. It had been sailed out from England via Panama by a young couple, Rosie and Colin Swale, and their delightful young children, Eve and James. *Annaliese* was an inshore catamaran, designed for sailing on the Thames Estuary. It was heavy, with twin diesel engines, very low bridgedeck clearance, narrow beam, and huge windows that flexed when you pushed against them. It seemed astonishing to me that it had survived the voyage out to Australia, despite sticking to the tropical trade winds.

Alarmingly, Rosie and Colin announced they had been inspired by David's circumnavigation in *Rehu Moana*, when he'd passed through the Magellan Straits in Patagonia, and now planned to return to England by running down their easting in the Pacific Ocean and doubling Cape Horn. They did, too. The book Rosie later wrote, *Children of Cape Horn,* echoed the title of David's book, *Children of Three Oceans.*

Colin was a Londoner, about 30, with curly brown hair, an unassuming air, and sardonic wit. Before he met Rosie, his ambition was to become a millionaire by the time he was 30, building up his lawn-mowing empire. He'd been well on track until he fell in love with Rosie. At times he seemed out of place in this madcap adventure, very much in the shadow of his extraordinary wife.

Rosie was an Irish urchin, who'd had a most irregular childhood, recounted in her book, *Rosie Darling,* and had already survived numerous adventures before Colin fell for her. She was a whimsical character, with long, blonde hair, very long legs, and an angelic face. In between fending off pirates and unwanted admirers in far-flung places, she'd worked as a model and knew how to seduce the camera. She had a smile that David could have used to melt icebergs.

David, not unnaturally, felt some responsibility for the situation. One night, after we talked it over, he decided to be frank with them. The four of us, plus the boisterous kids, had dinner together in *Annaliese's* crowded cabin, surrounded by nappies soaking in buckets, toys, and dismantled equipment. Inevitably, the conversation got around to their impending voyage. David told them that their catamaran would be matchwood in a serious Southern Ocean gale, imploring them to sail home in the trade winds. I thought even that would be a challenge.

Rosie and Colin were unfazed, so David gave them his *Gibson Girl* emergency transmitter, a device that sent out a distress signal when you cranked its handle. That turned out to be ironic, given that they completed their passage without incident, while he, having sailed much earlier, found himself struggling for survival in fierce spring storms. Since then, *Annaliese* has made another voyage from England to Australia via Panama, with new owners, and is currently moored near me in Mooloolaba, Queensland.

*Annaliese today, moored on the same walkway at Kawana Waters Marina in SE Queensland as my current yacht, Mehitabel. I walk past it every day.*

Sometimes we took *Ice Bird* out for sea-trials, usually with Susie and Vicki, David's daughters, who were now on the brink of adolescence. They'd be embarrassed by it now, and I don't blame them, but they both had a crush on me, as young girls do, and David had to rescue me from their clutches on more than one occasion.

One or another of David's girlfriends (he had at least two as far as I could tell) was also around most of the time, and we made a happy, if somewhat unconventional, family, all crowded into *Ice Bird's* small cabin. Sailing around the harbour was fun, and I began to dream once more of owning a yacht. For the first time since I arrived in Sydney, I began to think about my future, though the sense of alienation, the profound loss of something I could not fathom, remained.

On one of these sails, it began to rain. Wearing a wet spray jacket I'd borrowed from David, I leaned against the main bulkhead, on which someone had painted a large kangaroo and a kiwi. To my dismay, the paint smudged, the colours running down in a long streak. It was the sort of stupid thing I was always doing.

*The kangaroo screwing the kiwi, while a curious penguin looks on. This painting has now been restored by the Powerhouse Museum, as part of their project to preserve Ice Bird.*

David, who'd grown up in New Zealand but was now based in Australia, had been pleased with the painting, although he did remark that it looked like the kangaroo was screwing the kiwi, adding that it was an apt depiction of the trans-Tasman relationship. Now, when he saw the damage that I'd inflicted, he laughed, saying, 'The kiwi and the kangaroo have decided to get together and piss on the springbok.'

David had an irreverent sense of humour and an unorthodox social outlook. It was one of the things I loved most about him, besides his enthusiasm for life and his courage. His iceberg-blue eyes shone with the passion of an adolescent, or gleamed with mischief, something he never lost until the day he died, aged 85, in 2002. It was this passion, I believe, that drew people to him.

*David at the helm during a trial sail in the rain. The rest of us are hiding in the cabin. Photo: Fairfax Media.*

However, he was the most unsympathetic doctor I ever met. People seeking attention and sympathy, who account for many doctors' appointments, got short shrift. He was great in an emergency, though, which was amply demonstrated during World War Two, when he served with distinction as a medical paratrooper in France, and he admired people with pluck. He loved telling the story of the 14-year-old boy who was bitten by a snake while fishing. The boy killed the snake, cut it up, baited his lines with its body, pocketed the head for identification, then calmly rode his bike to David's surgery. Luckily, the snake was not deadly.

I remember the day that David showed me a letter he'd just received from his old friend, Bill Tilman, or to give him his full title, Major Harold William Tilman, CBE, DSO, MC and Bar. Tilman's exploits are legendary. Having wrecked two Bristol Channel pilot cutters, *Mischief* and *Seabreeze*, both victims of ice in Greenland, he had just bought another for further Arctic voyaging.

He wrote: *It is becoming difficult to find a good pilot cutter, I myself, of course, having accounted for two of them. I have just bought another, Baroque, though she is in rather poor shape. The surveyor, after a thorough examination of the ship, turned to me and said, 'Mr Tilman, the only thing holding the two sides of this vessel together is habit.' I have, nonetheless, bought her, and hope to make another attempt on Scoresby Sound next summer.*

David also gave me an introduction to Blondie Hasler, and I exchanged several letters with him. I was keen, of course, to get myself into a junk-rigged yacht, and mentioned my desire to build or buy the smallest boat possible, to attempt a singlehanded tropical trade wind circumnavigation.

Blondie sent me preliminary drawings of a 14' 5" flush-decked yacht with a single junk sail, called *Loner*. It had a very deep keel and, of course, no engine. He'd had the prototype built in fibreglass foam-sandwich composite by an expert in the field, Derek Kelsall, whom he knew from the 1964 OSTAR, but didn't have construction drawings, which I'd have to arrange for myself. One thing I liked was that the design included positive foam buoyancy and was unsinkable.

Blondie concluded by asking me which way around Australia I would go, north or south, on my way into the Indian Ocean. The idea of a 20-year-old building a boat like *Loner*, and setting off around the world, seemed perfectly reasonable to him. I think it *might* have been a good boat for me at that stage of life, or possibly the 20' version he designed soon after, but a month or two later I was talked out of it by well-meaning new friends.

The well-known Australian sailor, navigator and writer, Charles W Ure, commissioned a 20' version of Loner, which was built in New Zealand by Derek Kellsall, Blondie Hasler's friend and fellow OSTAR competitor. Charles Ure shipped it to England, then set off for Australia, via Panama, along the tradewind route, and sailed into oblivion. Despite the boat being unsinkable, no wreckage or trace of Mr Ure were ever found.

*The drawing that Blondie sent me of Loner, his 14' 5" miniature ocean voyager.*

The best party comes to an end eventually. The finale came right on cue, on 19 October 1972. That's the date David set for departure, and when the time came, we packed up the tools, stowed the provisions, and cast off the dock-lines. As a parting gift, David gave me a copy of his book, *We the Navigators,* in which he wrote:

*Without your friendship and help, the more southerly navigation now pending would scarcely have been possible.*

He also gave me his last remaining dollars, ensuring that two final cheques he'd written for suppliers bounced. The long-suffering Tim Curnow honoured them.

*David talking to Sylvia Cook on the foredeck of Ice Bird while John Fairfax looks on. Sylvia and John had just completed the first crossing of the Pacific Ocean under oars. Photo: Fairfax Media.*

The dock was crowded with well-wishers, TV cameras, Susie and Vicki, girlfriends, and celebrities, including John Fairfax and Sylvia Cook, who had just rowed across the Pacific. I felt a deep sadness that it was all over, especially in the light of my current circumstances.

It was a bright, spring day, clear and fresh, and Sydney Harbour was at its beautiful best, like one of those gigantic Brett Whitely paintings of the harbour, in which you are almost swallowed by the vibrant blue canvas.

I sailed *Ice Bird* down the harbour, tacking against the NE seabreeze, while David was interviewed by an ABC TV crew, who were perched uncomfortably on the sidedeck, camera, tripod and all. They had to change sides every time we tacked, amid much grumbling.

We were surrounded by a farewell flotilla, including the barque, *New Endeavour*. I was excited to see this ship, as it had featured in the *Sandefjord* movie. *Sandefjord*, I had discovered to my surprise, was well-remembered in Rushcutters Bay. It seemed like ancient history to me then, but almost everybody I talked to on the dock had a story about the boat and its colourful crew. I realise now that *Sandefjord's* visit had taken place just six years earlier. Six years is a much longer time when you are 20 than it is when you are 72. As always, the shifting sands of perception trick the mind.

*New Endeavour accompanies Ice Bird down Sydney Harbour on 19 October 1972.*

Each time the ABC TV crew stopped to replace a film or tape, David hurriedly turned to me, and I tried to show him how to adjust the self-steering gear. We'd only had one opportunity to trial it prior to departure. It was a bit like the blind leading the blind, but at least I had read the instruction manual. It was a servo-pendulum type, designed by Blondie Hasler and professionally built by Gibbs, but it was their smallest model and undersized for the boat. It had been the only unit available in Sydney, but did not cope with the Southern Ocean.

We'd originally fitted a homemade self-steering gear designed and built by Jos Doel, the proprietor of the Rushcutters Bay Chandlery and a former cruising sailor. The design had worked well on his own boat, and probably would have on *Ice Bird*, but we did not have the time to trial it properly. Instead, I inherited it. All I needed now was a boat to put it on.

Sometimes, while I was trying to demonstrate how the self-steering worked, other media representatives would zoom up and I had to lie down behind the spray dodgers flanking the cockpit, so David could be photographed sailing off *alone*. He'd clutch the tiller briefly and strike a heroic pose, trying to turn his grimace into a smile. He was a life poet, writing his poem on the public record, creating a dazzling story, but his innate character tended towards shyness, and he was beginning to show the strain.

On one occasion, the ABC TV crew were so engrossed with their task that they didn't notice *Ice Bird* approaching the rocky foreshore. If I just tacked, I was afraid they'd topple into the sea, camera, tripod and all. Subtle attempts to attract their attention didn't work, so I stood up. They frantically motioned me to get down. Just before disaster was unavoidable, I shouted, *tacking now*, and put the helm down. The ensuing scramble was entertaining to watch.

When we passed between North and South Heads, and the bows began to lift to the ocean swell, Tim Curnow rowed over from our escort vessel and took the TV crew and me off.

*Ice Bird rises to meet the sea at Sydney Heads, 19 October 1972, at the beginning of David's historic voyage to Antarctica. David waves farewell, as his literary agent, Tim Curnow, rows the ABC media crew and me across to our escort vessel.*

*Ice Bird* forged seawards, heeling to the strengthening NE seabreeze, swooping over the waves and flinging spray aside. The boat looked to be in its element at last, and David busied himself trimming sails. The circus was over.

I watched for as long as I could, as the little blue boat gradually merged with sea and sky. Then our host vessel turned back and I returned reluctantly to the city. I was thinking about David out there, alone at sea. Was he feeling lonely, too? I had his answer some years later when I read his book, *Ice Bird*. He could have been thinking of me when he wrote:

*I was not in the least lonely. This solitude was a different thing altogether from the lonely emptiness you suffer in a strange city where, knowing no one, you are surrounded by uncaring men and women, all supported by their own human ties.*

David's voyage to the Antarctic Peninsular is the stuff of legend now. After a brief stop in New Zealand to sort out a recalcitrant HF radio (it never worked), he pressed on. Weeks turned to months, and the silence became ominous. I rang Tim Curnow several times but after three months he told me that we had to consider David lost. He was more than six weeks overdue at his destination, the American Antarctic base, Palmer Station.

I remember taking the bus to Watsons Bay and going up onto the cliffs of South Head. I stared out to sea, thinking of the day that *Ice Bird* left, and how I'd watched the little blue boat heading east into the unknown. I wondered what David's last hours had been like, and my heart was sore. If that had been the end of the story, it might have changed the trajectory of my life.

Then I heard from Tim that David had arrived at Palmer Station after a prolonged struggle to survive, following a major storm and several capsizes. *Ice Bird* had been dismasted and severely flooded, and in the effort to bail the water out, David had suffered frostbitten fingers. It was over 14 weeks since he'd left. The images that came back were shocking. With his matted hair and haunted face, David looked like he'd been locked in a dungeon for the previous three months.

When I look at photos of the jury rigs David constructed on that voyage, two thoughts strike me. One is how ingenious they are, and the second is, how the hell did he erect them alone in 30' seas? It is hard enough to work on the deck of a yacht at sea in moderate weather, and often a real struggle in a gale, but a dismasted yacht has the motion of a mechanical bull. David sometimes came across as technically challenged, a dangerous man with a chisel in his hand, but he was the world's greatest jury-rigger.

He left *Ice Bird* at Palmer Station for the winter, travelling first to Washington and then to the South Pacific, where he worked with *National Geographic* magazine on an article about traditional Polynesian navigators. They also contracted him to produce two articles about his Antarctic adventures. When he returned to Sydney, we had a brief but happy reunion, and he introduced me to his son, Barry, and Barry's girlfriend, Ros Dawes.

David returned to Palmer Station the following summer, where he managed, with the help of willing Station personnel, to re-rig the boat as a gaff cutter, using an old spar from the base and a re-cut mainsail. He replaced the smashed Hasler self-steering gear with an Aries unit, more robust in design and better suited to the size of the vessel. He also had some decent polar clothing, courtesy of *National Geographic* magazine. He then set off to explore the peninsular, pushing further south than any small vessel had ever been.

*David pushed further south into the ice than any yacht had ever been. Photo: David Lewis archives, courtesy Barry Lewis.*

After departing the Antarctic for Australia, another storm capsized *Ice Bird* in the South Atlantic, dismasting the yacht again and damaging the self-steering gear. David was struck on the head by the boom, which left him with detached retinas that would later come back to haunt him. He concocted another magnificent jury rig and headed for Cape Town, where he left the boat in March 1974. He had other commitments, though he freely admitted he'd also had enough of the voyage, at least for the time being.

Barry Lewis flew to Cape Town and re-rigged *Ice Bird* yet again, with a heavy timber mast he sourced there, and galvanised wire standing rigging. He then brought the boat home across the Indian Ocean, known as the roughest of the Southern Ocean passages. He sailed conservatively, reefing early and towing warps when needed, and 86 days later he came in through Sydney Heads after a passage of 6000 miles. It was a magnificent display of seamanship, and he had no major problems, despite some F9-10 winds at times, though he was a lot further north than David had been when overwhelmed.

*Ice Bird approaching Cape Town from Palmer Station, Antarctica, under yet another magnificent jury rig. Photo: David Lewis archives, courtesy Barry Lewis.*

After David sailed out of Sydney that October afternoon in 1972, one of his girlfriends drove me back to my room in Surry Hills in her van. When we stopped outside my building, she unexpectedly hugged me and suggested I come to Canberra with her. I leapt out of the van, rather ungraciously, like a startled cat, flushed with shame. There wasn't any ocean there anyway, I told myself, and by all accounts it was bloody freezing in winter.

A wealthy, middle-aged patron of David's then offered me work on his yacht, and a bunk in the boathouse below his Rose Bay mansion, which may have been fun, but his young wife treated me as the opposition, so I quickly moved on, taking up Rosie and Colin's offer to work on *Annaliese*.

*Annaliese's* cabin was largely made up of windows, as mentioned earlier. We bolted plywood over the windows and fibreglassed half-round battens to the inside of the narrow vertical areas between them. This strengthened the cabin somewhat but made it gloomy.

Sometimes, when things were going wrong, Colin became rather gloomy himself, and I found it easier to stay out of his way. On these occasions, I took little Eve and James across to Rushcutters Bay Park and played games with them. Soon they were clamouring for my attention, and my heart swelled with affection. I have always loved children (when they are happy), with their bright eyes and radiant smiles.

One day I discovered that Colin and Rosie had crossed tacks with *Kelasa* in Barbados. Adrienne and Harry, who had not seen *Otaha* before leaving Durban, but knew I planned to chase them in my small catamaran, had rowed over to *Annaliese* on the off-chance it might be me. I sighed at such touching faith.

A few days later, Rosie excitedly waved a letter at me. It was from Adrienne, who was now back in Australia, living in Brisbane. She'd seen an article about Colin and Rosie in a local magazine and written to them care of the editor. I had to read the letter twice to realise that Adrienne was just up the coast. What an extraordinary thing chance is. First Brett, who may have saved my life, then David, who almost certainly did, then Rosie and Colin, and now Adrienne.

I immediately wrote to her, and was thrilled to receive a long reply within days. *Kelasa* had sailed from Durban to Honolulu in just nine months. Adrienne had been disappointed to miss out on cruising the West Indies, but the main reason she'd jumped ship, and the reason why Harry had been in a rush to get to Honolulu, was that he'd decided to sell *Kelasa* and build a new, faster boat. Adrienne, of course, was expected to earn the money, while he sharpened his tools.

One day when I was working alone on *Annaliese*, a local yachtsman stopped by and told me that he thought Colin and Rosie were braver than Ned Kelly, which amused them when I mentioned it later, though it was true. Without detracting from their magnificent voyage, they were lucky with the weather. Bill King, who made the same passage that season on his junk-rigged schooner, *Galway Blazer II,* reported weeks of benign, sunny days en route to Cape Horn, when he could sunbathe in the cockpit. I am sure that if *Annaliese* had met a serious Southern Ocean storm, the boat would not have survived.

By the time they left, Jos Doel had introduced me to a multi-millionaire who was looking for a paid hand on his motor yacht. This man, with a reputation for being a tough businessman, always treated me with civility. All I had to do was polish the rails, touch up the varnish, and crew for him, in return for a decent wage. I was secure at last, though it hardly felt that way. Sydney still seemed like a monster that might swallow me.

# Chapter Nine

# At the Gates of the Pantheon

Luckily, I soon made new friends. One was a German shipwright, Rudy Krause, who'd built his family a stunning, canoe-sterned, carvel-planked yacht, *Rainbow*, with sweet sheer and varnished topsides, designed by William Garden. He'd gone cruising with his wife, Olga, and two teenage daughters, but they were back now to give the girls an education, living on the boat in Rushcutters Bay. Rudy still hankered for the sea, and we spent many happy hours talking about ocean voyaging. Olga was more reserved, especially when her daughters were around. I think she was protecting them, given that I was a young, single guy, and a bit footloose.

In January 1973, I moved into the spare bedroom of another new friend, Geoff Collins, whom I'd met through Gavin Fox, whom I'd met through Colin and Rosie. Colin's father had been Gavin's headmaster. Colin and Gavin had hardly known each other as boys, since Colin was sent to a different school, but Gavin recognised him when reading a newspaper article about *Annaliese*. And thus the world turns, as Bernard Moitessier once said.

In two shakes of a stick, I'd started building another Wharram catamaran, to the same design as *Otaha*. This time I was going to make a proper job of it, and I had the assistance of Gavin and Geoff to achieve my goal. They were Yorkshire men, a decade older than me, and we became close friends. I always felt comfortable with older men in my youth, seeking a father-figure perhaps. Guys my own age often made me nervous. Gavin was a charismatic, full-faced, ginger-bearded clown, who danced on tables with umbrellas when he had a drink under his belt, while Geoff was dark-haired, fine-boned and introspective. They both had that dogged quality so common in Yorkshiremen, and the project came together well under their steady influence.

*Setting up the hulls of my new Wharram Tane catamaran. I am at the bow while Gavin Fox lines up a stringer at the stern. Photo: Geoff Collins.*

We were building the boat in the front yard of Gavin's friends, Jock and Dee, in Parawi Road, Mosman. As I walked down to the property, I could look out through gaps between the houses and see the rugged, stratified, sandstone cliffs of Sydney Heads jutting out into the Pacific Ocean. The orange and grey cliffs contrasted pleasantly with the ocean, which on good days looked like a huge, bright-blue prairie, stretching to infinity. I remembered Harry Pidgeon's words that the sea was a great blue highway that would take him anywhere he wanted to go.

It was a sight that filled me with longing to escape. *The sea, the sea.* Sydney Harbour had lost none of its charm, but I was frightened of the city, and became impatient with the slow progress of the catamaran's construction. I couldn't explain this to my new friends with any emotional honesty, though couched in the romantic terms of sea fever it was acceptable.

*The hulls were coming along nicely under the influence of Gavin and Geoff, seen here in the white jumper. Photo: Gavin Fox.*

Another reason for my impatience was an infatuation with Cammeray Marina, where new friends, Daniel and Nicole Moulin, were fitting out their 35' sloop, *Henri*. They were the young French couple that Henri Cordovero, on the yacht *Challenge*, had told me about in Mauritius. They'd named their boat in his honour. I had forgotten about them during those first harrowing weeks in Sydney, and then been too busy once I got involved with David. One day, shortly after David sailed, and when Colin was in one of his grumpy moods, I dug the note that Henri gave me out of my wallet and looked at it. It was in French, but I understood the address.

It had been a cold day, with weak sunshine and a razor-sharp wind. I'd been depressed and could hardly afford the bus fare, but nonetheless decided to make the effort. The bus left from the ironic-sounding Spring Street, near Circular Quay. I eventually found the right bus stop and huddled in it, bogged in misery.

After a while, I noticed a small tree nearby, a sapling really, that seemed to be planted in a crack in the pavement. When buses stopped, they belched diesel fumes over it, and the wind seemed to be trying to tear it out of the ground. People brushed impatiently past, threatening to trample it, but it looked remarkably healthy, as it swayed and bobbed in the breeze, sunlight flashing on its silver-backed leaves. It looked like it was dancing.

When my bus eventually came, I took something of that tree's *joie de vivre* with me. My sense of spiritual renewal was heightened as the bus made its way over the Sydney Harbour

Bridge, where I could look out past the Opera House, which was nearing completion, at the bewitching harbour, and was further strengthened by my discovery of the leafy, relaxed suburbs of the affluent north shore. This was a very different environment from the city centre, and it enchanted me.

At Cammeray Marina, in Long Bay, Middle Harbour, surrounded by steep, bush-clad hills, I walked diffidently along the dock, feeling like an intruder, or refugee. Surely, they could see I had no right to be there. Then I found *Henri*. It was a pretty boat, with classic lines and a sweeping sheer. I later found out it was a Lion class, designed by the English naval architect, Arthur Robb, and built with strip-planked timber. A group of people were working industriously aboard. When somebody walked by and shouted out an invitation for a beer, a handsome young man with a French accent shouted back, 'Too busy, thanks all the same.'

I just stood there, too shy to say anything. I was about to walk away when the man noticed me and said brusquely, 'What do you want?' I diffidently handed over the note. He read it with a frown, and then broke into a huge grin. 'Nicole, Nicole,' he shouted, waving the note in the air, a letter from Henri.' He turned to me, 'Come aboard, come aboard, I am Daniel.' It was the beginning of a beautiful friendship.

Nicole was a short, slim girl, with unruly black curls, a sharp mind and even sharper chisels. Besides being a superb carpenter, she had prodigious artistic talent, and was transforming *Henri* into a unique, seaworthy home. There was nothing I loved better than spending a few hours on aboard *Henri* with Daniel, Nicole and their young daughters, Celine and Garlun, soaking up the atmosphere of Cammeray Marina. It was almost like being back in Durban.

The proprietors, Bunny (Ian) and Fran Rabbitts, had once been ocean voyagers, and advertised the marina in *Pacific Islands Monthly*, which in those days had a section on voyaging yachts and their whereabouts. Ocean voyagers read it to find out where their friends were, and there were always foreign-flagged yachts spending the summer months at Cammeray Marina. That's how things worked in the days before the internet.

I got to know a few of these visiting yachts and their people, in particular, Scott and Kitty Kuhner aboard American-flagged *Bebinka*, an Allied Seawind 30 ketch, a very pretty and solidly-built boat. I'd previously seen another Allied Seawind 30, *Apogee*, in Durban in 1967, sailed solo by Alan Eddy, which became the first fibreglass yacht to circumnavigate the world. Scott and Kitty were only in their 20s, full of youthful fun, and yet they had a great feeling for the traditions of ocean voyaging. They'd been influenced by the Hiscocks, having read Eric's book, *Around the World in Wanderer III.*

They set up their twin jibs in the same way the Hiscocks did, with the poles hinging upwards to nestle in chocks at the spreaders, ready for instant deployment, and they also carried twin storm jibs, a trysail, a mizzen staysail, and the usual assortment of headsails. *Bebinka* was amazingly fast, averaging 125nm a day in the trade winds, often exceeding this considerably in fresh conditions. Interestingly, they said that *Bebinka* did not roll too badly under twin jibs, unlike *Wanderer III*, which Eric Hiscock complained about in letters and his books.

Before leaving the USA, Scott and Kitty changed their alcohol (metho) stove for a kerosene one, fitted a Hasler wind-driven self-steering gear, and replaced the petrol engine with a 20hp diesel motor, which could be hand-cranked when battery power was low. They used kerosene lamps below, to reduce electrical consumption, and ran in 'dark ship' mode in the lonelier stretches of ocean. There was no refrigeration. In New Zealand, they replaced *Bebinka's* 10-year-old standing rigging with galvanised wire. I was interested to hear that they'd met the Hiscocks in New Zealand, and that Eric had complained about 49' *Wanderer IV* being too big. In recent years, Scott and Kitty have written a book about *Bebinka's* world voyage, *The Voyage of Bebinka.*

I began to long for the day when I could join them, and this growing impatience led to my downfall. It all started innocently enough. At a coffee shop in Rushcutters Bay, I met a man called Gordon, a slim, elegant old bohemian, who still wore the tight trousers and black skivvies of his youth, and despised bourgeoisie society. He didn't talk to many people, but took a shine to me because I was a mere boy, and listened to his stories with eager interest.

When the bohemians were driven out of Kings Cross by petty criminals who wanted to take over their real estate, Gordon retreated to a small, double-ended timber ketch, with larch planking on English oak timbers, called *Simba*. The boat lay on a mooring in nearby in Elizabeth Bay, and looked, to my eye, a little like a miniature version of *Sandefjord*. There, Gordon dreamed of sailing away on a grand adventure, but like his earlier dreams of artistic independence, they came to nothing.

*Simba* was the magic length, 24', the same as *Dove*, and as cute a boat as you ever saw, with a black hull, cream decks, and short, stout, varnished masts. There were ratlines up the rigging, red and green side-screens to port and starboard for kerosene navigation lights, solid rails around the decks, two anchors stowed on the foredeck, and even little storm shutters that folded down over the portholes. I was always sighing over the boat.

One day, out of the blue, Gordon said, 'I want you to buy *Simba*.'

'But I can't. I've got a half-built catamaran and no money.'

'I'm getting too old. I can't manage the maintenance any more, and sooner or later one of those damned speedboats is going to run me down while I'm rowing ashore. I'll sell it to you for one thousand dollars.'

I went home with my head spinning. If only I had the money, I could buy *Simba,* take it around to Cammeray Marina and live aboard right *now.* I resolved to borrow the money from friends.

It never occurred to me that I had an obligation to Jock and Dee, who'd offered me their backyard, at least in part, to provide an example to their 13-year-old son of how dreams could be fulfilled. Not to mention all the free labour Gavin and Geoff had put into the project. Geoff even forked out half the money for *Simba*. For his reward, he immediately had to look for a new flat-mate.

The day I finally took possession of *Simba,* I thought I'd died and gone to heaven. The boat was perfect. True, it had no headroom, being a converted navy whaler, unless you were five years old. Celine and Garlun loved it, they thought it was a floating doll's house. *Simba* had no engine, no radio, no electrics, no tanks, no liferaft, etc. I had no intention of fitting them either. John Guzzwell had none of those things on *Trekka*. Well, he did have a small, disreputable-looking Seagull outboard motor, rarely used. There were two bunks amidships, a few buckets, a couple of kerosene lamps, and a one-burner kerosene stove. What more could any self-respecting vagabond want?

*I will never be unhappy again*, I told myself, as I hurried back down to *Simba* with the ownership papers in my hands. As I walked down the steep hill to Elizabeth Bay, I thought I heard distant thunder, but suspect it was the gods laughing. I quickly untied *Simba* and sailed it around to Cammeray Marina. As mentioned before, I'd fallen in love with this marina. It had only been established in 1965, but resembled the old boatsheds that were so common around Sydney Harbour foreshores when Port Jackson was a working port.

When I arrived in 1972, the rightly-famous Charlie Busch's Boatshed in Rushcutters Bay was still standing, but had been taken over by Ron D'Albora, who had built a marina there. Charlie Busch's Boatshed was where *Kelasa* had taken a mooring, and where

Adrienne Matzenik had kept her old cutter, *Dorothea Mackellar,* (formerly and currently known as *Utiekah II*, originally built for the famous Harold Nossiter Snr).

The colourful characters who had congregated at Charlie's, renting moorings from him, convening in the shed around the dinghy racks, and cruising in company around the harbour, had long dispersed to who knows where. Not long after, the historic boatshed was demolished, to be replaced by another of the ubiquitous brick and tile buildings that marked progress in that era. The sort of sacrilege an old curmudgeon like me might call a blot on the landscape.

It was the start of a trend, another cultural shift, but in the early 1970s there were still a few of the old boatsheds left. One I remember with particular fondness was not far from Cammeray Marina, Tom Joel's place. It was a ramshackle establishment, with a small, weatherbeaten timber cottage that Tom lived in, a rickety jetty, and an old slipway that could haul boats of 30' or less. It was a very laid-back establishment. Even then, it was surrounded by expensive, multi-story houses that glowered down on the well-organised chaos below.

Tom was a short, hirsute man with an unprepossessing appearance, but you could tell immediately on meeting him that he would never gouge you financially. In fact, he had a soft spot for impecunious battlers, and the jetty often had project boats rafted alongside. I loved rowing over there in my dinghy and yarning with him. Sometimes I had a little job I wanted to do. The loss of this maritime heritage to monotonous, up-market real estate has impoverished Sydney Harbour, in my opinion.

Another person I remember with affection was Bob Gordon, who was building boats in Lavender Bay when I met him, but had earlier been based in Berrys Bay, a little further up the harbour. He was another of the great waterfront characters of Sydney when it was a working port. I went looking for him after seeing one of his yachts, *Widgeon,* on the Cruising Yacht Club slipway in 1973.

*Widgeon* was a strikingly handsome yacht, and I was not surprised to discover that the 42' carvel-planked hull had been built to the lines of Conor O'Brien's *Saoirse,* the first yacht to circumnavigate the world via the Southern Ocean, between 1923-5. Bob had replaced the original design's gaff rig with a bermudan ketch rig inspired by L. F. Herreschoff. Bob had a thing for Herreschoff designs, and built a number of them over the years. I reasoned that a man who built boats like *Widgeon* would be a kindred spirit, and although I might have inflated my own net worth, I certainly never underestimated his.

*Widgeon sailing out of Sydney Harbour, bound for Lord Howe Island. Photo: courtesy Robert Gordon.*

Bob was an excellent shipwright, and also a great raconteur. He went his idiosyncratic way with an extraordinary level of self-belief, and an inexhaustible amount of energy. At the age of 16, he gave up working as a shipwright's apprentice after just three months, deciding he'd learned enough to get along on his own. It sounds like teenage hyperbole, but he proved his case by immediately laying the keel of his first boat, a 30' John Hanna Tahiti Ketch, and went on to build more than 30 substantial vessels over the next 67 years.

I doubt Bob made much money from his lifetime vocation. At one stage, he and his wife lived in a humpy alongside the boats he was building in Berrys Bay, and he also engaged in commercial fishing to pay bills. He didn't seem to seek commissions to build boats, just built the vessels he wanted to, the way he wanted, and trusted that he'd find a buyer for them in due course, which he always did.

I have never known anybody who worked so hard, but he was also generous with his time and ideas. What I remember best about him was his warmth and humanity. You could see it in his wonderful, genial face, and he was full of amusing aphorisms that kept me laughing. I found him an endearingly unvarnished character. He created the life he wanted to lead and made no attempt to conform to societal norms, although he had such an innately decent character that he was more an example of integrity than rebel without a cause.

I remember the day we went over to Berrys Bay in Bob's old car to visit a mutual friend who was building a boat in a shed there. The car had no brakes, it seemed to me, or very poor ones, and on the downhill runs we scraped through intersections in first gear. I was a bit nervous, but Bob was perfectly relaxed, carrying on a wonderful narrative of boats and their people. I loved his stories.

In the early 1980s, when I mentioned to Bob that David Lewis was looking for an expedition ship to return to the Antarctic, Bob showed me *Tunny*, his 65' steel interpretation of L. F. Herreschoff's Marco Polo design, which was lying on a mooring in Lavender Bay. David and the Oceanic Research Foundation subsequently bought *Tunny* and transformed it into the three-masted *Dick Smith Explorer*, which made three successful voyages to the Antarctic.

The last time I saw Bob was in the 1990s, when I sailed off permanently to Queensland, but his influence remains with me. By that time, the developers had pushed him off the land he'd been using next to Neptune Slipway for many decades, and he'd retreated into the nearby tunnel under the railway viaduct that he'd previously used as a storage shed. He continued to build boats in the tunnel until his death in 2006 at the age of 84, working alongside his son, Robert, who is still building boats there.

As soon as I brought *Simba* to Cammeray Marina, I quit the dream job as deckhand on the millionaire's yacht and went to work in a factory. My reasoning was that the factory was closer to the marina, and anyway, Daniel worked there. Unfortunately, Daniel quit a week later, having secured a better job elsewhere, and I discovered that factories were not much fun. I was to work in many factories over the next decade, but never lasted in any of them for longer than three months.

Working at some unskilled job for three months, long enough to restock the boat with baked beans, had always been the original plan, but I was beginning to discover that this idea had a few holes in it. Some of them were in the boat, speaking literally, because you could never earn enough to get a decent vessel while working for minimum wages. Despite realising this, I was still too obsessed with the voyaging dream to change tactics.

*Poeme on a mooring at Cammeray Marina, January 1974.*

On this occasion, I was hounded by an emaciated alcoholic who looked like he never took a bath, and who always had a dog-eared, self-rolled cigarette hanging out of the corner of his mouth. His name was Dick, but the other workers called him *Suck My Dick,* because that was his favourite expression. Once, during lunch, I was reading the *Sydney Morning Herald*, a quality broadsheet newspaper, when it was snatched from my hands and thrown to the ground.

'You won't find no sex in that,' said *Suck My Dick,* thrusting a crumpled magazine at me. It had a lurid title, like *Big Tits* or something, and appeared to be covered in greasy stains. I leapt away in alarm, causing the rest of the men to guffaw delightedly. *Suck My Dick* leered at me. 'Got a girlfriend, have you? Ever seen a fanny, sunshine?' I blushed furiously, causing more guffaws.

I took to eating my lunch outdoors, even when it rained, but there was no escaping my tormentor during the shift, when we had to work side by side, shovelling wet chemical sludge out of vats onto trays, from where it was wheeled into large ovens to be baked. *Suck My Dick* took to pinching my backside whenever I had my arms full, while the other men laughed derisively. Within a few weeks, I was suicidal, but luckily Daniel saved my bacon by getting me a job at his new place.

On weekends I sometimes took *Simba* for a sail. The boat was more tender that I expected, given the short masts and conservative sail area, but apart from being under-canvassed in light winds, I was happy with its performance.

There were a few amusing incidents. Once, while I was returning up Middle Harbour with a friend, the wind died and a steady drizzle set in. I launched the dinghy and began towing *Simba* back to the mooring, about a mile or so in flat water. It was slow but satisfying. Shift with your own hands, as Moitessier said. Soon it was raining steadily, but was otherwise a warm summer day, and I enjoyed the experience despite being soaked.

I learned how to make the most effective use of my motive power. Stroke a few times easily on the oars, watching the towrope intently. As it straightens out, give a few decisive strokes on the oars, then ease off again as the boat follows you. Repeat. You can go on forever like that without getting tired, at least when you are 21.

My friend stood on deck, wearing my three-quarter length wet-weather jacket, my only waterproof garment. I was happy for him to have it, but he wasn't that happy, with water dripping off his nose, arms folded crossly over his chest, glowering at me. When I refused a tow, preferring, as I tried to explain later, the pleasure of being independent (like my heroes), his glowering darkened the sky.

On another occasion, coming in from the ocean with a light but steady easterly astern, I came up to the Spit Bridge, the gateway to inner Middle Harbour. I decided to sail under the bridge rather than wait for it to open, as I knew my masts were low enough to clear the central span, and I could avoid the rush of other vessels when the light went green. Anyway, you were not supposed to sail through the opened bridge, and I'd been lucky to get a tow going out that morning. I assumed that sailing through when it was closed, if you could, was permissible. I could see the bridge-master in his tower looking concerned as I approached, but I just gave him a little wave.

All went well until *Simba* came out the other side, when the wind, eddying around the vertical concrete bridge supports, became capricious and headed us. The sails came aback and the boat fell off, coming to rest gently against one of the timber bulwarks that protect the supports from collision damage. Having seen me go under the bridge, but not emerge on the other side, the bridge-master came down from his tower, crossed over four lanes o f traffic, and peered over the railing at me. Just then, a puff of easterly wind arrived, and I sailed serenely away, giving him a larrikin's grin.

One of the yachts at Cammeray Marina was a high-sided 30' ketch, *Rendezvous*, that I knew about from reading a small, battered, paperback book given to me by *Ellimata's* skipper, Curtis de Camp, in 1969, a cruising guide to Queensland called *Cruising the Coral Coast*, by Alan Lucas. The book had a formative influence on my decision to head for Australia in 1972. Alan had designed and built *Rendezvous* in Sydney in his early twenties, and cruised it extensively on the Queensland coast, as well as further afield in Papua New Guinea and Indonesia. He'd met Robin Lee Graham aboard *Dove* in the Torres Strait in 1967, and the year before he'd begun a lifelong friendship with Tom Corkill, when *Clipper* and *Rendezvous* crossed wakes.

*Rendezvous* was about to depart on a circumnavigation with new owners, Andy and Bob, and at their farewell party I met Alan Lucas for the first time. Alan then had a 30' steel catboat called *Tientos,* that he was using to survey the New South Wales coast for a cruising guide to New South Wales, and to update *Cruising the Coral Coast.* People joked that he knew every rock and sandbank on the coast because he'd hit them. This was, of course, a joke, but he was certainly fearless in exploring areas that most yachts would be loathe to approach. He was great fun, and played wonderful flamenco music on his 12-string guitar. I didn't get to know him well then, but in later years formed a rewarding friendship with both him and his wife, Patricia.

On New Year's Day, 1974, I drank a toast to *Simba,* and swore that next winter we'd be in the tropics, far away from Sydney. I liked living aboard in Cammeray Marina, but it was merely an oasis. Soon, autumn was looming. *Bebinka*, and the other voyaging yachts that had berthed in the marina for the summer, were taking off, but this time I was going too. I planned to sail *Simba* to New Zealand, then up to Tahiti. The hull was white now with black trim, more suitable for the tropics.

Only the boat wasn't called *Simba* any more. 'That's a stupid name for a boat,' Daniel said. 'Why not call her *Poeme*. After all, that's what she is, a little poem, almost too perfect in fact.' So, I changed the name to *Poeme,* and Nicole carved me a magnificent name board.

Looking back, it is hard to believe that I thought I could get away with it. I'd discovered a fair bit of rot in *Poeme* during the summer, among other problems. There were broken frames in the stern, a cracked sheer plank, worn rudder pintles, and I winced when the decks creaked underfoot. The boat needed a rebuild.

I should've stayed another year and fixed the problems, or at least hopped up the coast in fair weather and pottered around in Queensland, which was a sort of sheltered workshop for decrepit boats from what I'd heard, but unfortunately, I wasn't that smart. About the only thing I did, apart from careening and antifouling *Poeme* against the rusting hulk of an old sailing ship, *Itata*, in nearby Salt Pan Creek, was to bolt on the self-steering gear I'd inherited from David Lewis.

*Poeme careened alongside the old sailing ship hulk in Salt Pan Creek, opposite Cammeray Marina, February 1974. Photo: Daniel and Nicole Moulin. The lines off the main masthead are to keep Poeme upright when the tide goes out. I was a little dubious about the system, as the lines were not that taut, but the old-timer who advised me knew what he was doing, and the boat sat there as steady as a monument. I used this system many times in later years.*

I consoled myself with memories of all the other less-than-perfect boats that had crossed oceans, Lee Graham's *Dove*, and Julio Villar's *Mistral*, in particular. They both remained a beacon of what was possible. Besides, on 12 April I would be 22, If I waited until everything was perfect, I'd be an old man of 30 or something. I was desperate to join my heroes, those lesser gods who resided in the Pantheon of the High Seas, to warm my hands in theirs in the far-flung ports of paradise, after crossing the wide blue ocean under sa il.

A week or so before I left, some friends, Hugh and Beryl, suggested I take a young woman they knew as crew. Hugh was an art dealer, and he and Beryl were sophisticated, charming people. They were building a huge, ferro-cement yacht called *Wedgewood* on a vacant block of land in Crows Nest in North Sydney. You could do that sort of thing around Sydney in those days. Fat chance now, amateur boatbuilders have been squeezed out by councils and developers. The boat had a Wedgwood-blue hull with white motifs, and the interior had alcoves in the ferro-cement bulkheads in which they had lodged white Greek figurines. It was quite the thing.

What could I say? Certainly not that I wasn't interested. One of the few advantages of being old is that people have stopped trying to marry me off. Not that I wouldn't have *liked* a partner like Patti Graham, but it just wasn't possible, for reasons I couldn't really explain to myself, let alone anyone else. A dinner party was arranged. Colleen O'Brien was attractive, with a cherubic face framed by a mass of brown curls, and lightly-tanned skin that glowed. She seemed interested, though whether in me, the adventure, or both, I wasn't sure.

'You'll have to come and look at the boat,' I said after dinner, thinking about the cramped cabin, the overflowing garbage bag in the cockpit, and the dirty laundry on my bunk. I was quite sure she'd refuse. I might even have neglected my housekeeping a little more than usual.

We all traipsed down to the boat after dinner, despite the fact that a cold drizzle had set in, as it does so often in Sydney. Hugh and Beryl had not seen *Poeme* before, and it was obvious from the look on their faces that they just wanted to pick up their pretty young friend and run. Colleen, however, dove straight into the cabin. She came back out a few minutes later, grinning. 'Yes,' she said, 'I'll come.'

Hugh and Beryl threw a farewell party alongside the hull of *Wedgwood*, with a bonfire and an old bathtub full of ice and beer. Everybody was there, even Jock and Dee, though they hadn't quite forgiven me for abandoning the catamaran in their front yard. I had only just sold the hulls to a friend of David Lewis, a young anthropologist called Leith Duncan, who'd trucked them to Melbourne. He later sailed the finished cat to New Zealand.

People put presents in a wheelbarrow, and it was soon overflowing. Among the tins of spaghetti and stew, there were some real gems; a sextant, a very expensive medical kit, and new wet-weather gear. One of the ironies of my life is that, despite my inability to form intimate relationships, and the unshakeable sense that I am trapped inside an invisible box, I have always been surrounded by wonderful friends. These gifts were gratefully received, as I had just $10 left in my pocket. I wasn't worried about finances, though, as this was in the best tradition of my heroes, those lesser gods of my imagination.

# Chapter Ten

# Any Which Way but East

On the following Saturday morning, 7 April 1974, my friends all came down to Cammeray Marina, threw streamers and sang *bon voyage*. There was no wind, a bit of a problem for an engineless boat, but I'd arranged a launch to tow *Poeme* out past Sydney Heads, since I wanted to get away from the coast before dark. This, I reasoned, was in the best square-rigger tradition. They'd often taken a pluck out to sea. I was feeling remarkably relaxed. Colleen seemed calm and cheerful too. The difference between us was that she would stay that way throughout the drama that was about to unfold. There are no photos in this chapter because initially I was too preoccupied to take any, and later the cameras were ruined.

Unfortunately, the driver of the launch was no sailor. Despite my angry gesticulations, he cast *Poeme* off under the towering cliffs of South Head, where the swell sucked hungrily on the rocks, and the tide carried us ever closer to destruction. I took out my big oar and began to scull, but it was hard to make progress against the long SE swell. I sweated and swore and wept with rage. Colleen graciously retired into the cabin.

It was 0200 before a westerly land-breeze filled in. If I'd had a small motor, I could have been 50 miles offshore by then. *One day I'm going to put an engine in this boat,* I said to myself. So much for the purists. The gods laughed.

Then the breeze freshened, and *Poeme* began to sail as fast as it ever would, given its dumpy rig, east across the Tasman Sea. By dawn there was no land in sight. We were surrounded by an endless succession of sloppy, grey waves, some of which slurped on board, and the sky was a ragged mess of tumbling clouds.

Having dozed all night in the cockpit, keeping an eye out for ships, I looked over the rail with a jaundiced eye. I knew I should take a sun sight and plot my position, but I didn't feel like teaching myself how to do it. I'd planned to familiarise myself with the theory before leaving but had been too busy. Anyway, lots of my heroes taught themselves to navigate as they went along.

I did take out the sextant, however, when Colleen wasn't looking, and gave it a puzzled stare. It was a strange device. I'd done something terrible in the days before departure. Short of cash, I'd sold the marine sextant I'd been given as a farewell present, and decided to use the old box sextant that came with the boat. This device had been designed for use in aeroplanes during World War Two, and had a bubble horizon. After a while, none the wiser, I put it back in its bag (it wasn't grand enough to have a box) and resumed my sullen watch.

For the first two days, the boat ran easily out to sea. The self-steering gear that had originally been made for *Ice Bird* steered *Poeme* in a desultory fashion, with the aid of some shock cord and frequent, vicious, kicks, a triumph of hope over logic. Lacking a differential linkage between vane and trim tab, it tended to over-steer. Colleen was unable to master the technique of this. She had no assistance from her increasingly gloomy skipper, so stayed below, passing up hot food at regular intervals. I stayed in the cockpit, dozing fitfully, opening an eye occasionally to stare balefully at the ocean and give the tiller another kick.

A strong SE gale built up on 9 April. *Poeme* lay ahull, broadside to the seas with no sail up. Colleen and I lay below, listening to the wind shrilling in the rig, and to the hiss of approaching seas. There would be a pause, a moment's silence, then the crest would crash aboard, water spurting in through several places. Occasionally, there was a louder hiss, once or twice a roar like an approaching train, followed by a much longer silence, then the boat was thrown violently aside, heeling until the leeward portholes were submerged, timbers groaning as water cascaded across the decks. We were pumping frequently, and our bunks were just wet squabs of foam. When we lay down, we lay in a puddle.

After a period when several crests in quick succession broke heavily aboard, I decided to try running before the storm, trailing a 300' x 5/8" diameter warp, given to me by Larry Bryant, an American friend I'd met in Cammeray Marina on his 62' gaff ketch, *Ron of*

*Argyll*. I discovered later that Larry was a psychologist, so perhaps I should have asked for some counselling as well.

After streaming the warp in a bight astern, I found the effect quite dramatic, but not in the way I'd hoped. *Poeme* continued lying ahull, but now the waves slammed aboard much harder, threatening imminent structural damage. I was aware that I needed to raise some sail forward, get the boat moving and allow the warp to drag the stern around. I could then sheet the storm jib flat, like Knox-Johnston did on *Suhaili*, and John Sowden on *Tarmin*. But that would have meant a trip onto the foredeck, to perch, high and unprotected on the dinghy, which dominated that space. *Poeme* was too small to stow the dinghy behind the mast, where it properly belonged.

It was easier to just pull the warp back in, although now I had a huge tangle of wet rope in the cockpit. I felt too tired and battered to stow it properly. Then, as I crouched in the cockpit, two nearby swells peaked simultaneously, producing one wave substantially bigger than those around it. Seconds later, it broke with a deafening roar, demolishing the last vestiges of my courage. Quickly tying the tiller to leeward, I scrambled below, into the cold, soggy sanctuary of my bunk.

There was only one thing I wanted to do now, return to the safety of Australia. Go home. My longing for the sea had mysteriously vanished. The gale, of course, was blowing us that way. It was only a matter of time before we arrived. I began staring at the western horizon with increasing frequency, but my gaze was always met by an endless vista of large, foam-backed waves. How far offshore were we? I had no idea. Maybe 100 miles by this stage, but it was just a stab in the dark. I'd literally lost the plot.

How long could this gale last? Surely it had to blow out soon? I would doze off, only to awaken as another wave crashed aboard. For a moment, with my sleep-dulled senses, I would be convinced that the wind had eased, then I would hear the rising shriek of another gust, and my heart plummeted. When I looked outside, my stomach churned at the sight of the waves, with crests falling down their faces, long white streaks of foam everywhere, and occasional large patches like snow drifts.

By the fourth day, I was becoming numb with despair. We were also soaked to the bone and freezing cold. That night, there was very little sleep. Surely the boat would crash onto the rocks at any moment? Shortly after dawn on the fifth day, we sighted land astern. We had obviously been much further offshore than I had estimated. I discovered later that there is often a fierce, east-setting current in this area.

I immediately got sail up, despite the continuing onshore gale, and began careering towards land. I knew from my reading that a good seaman would have been trying to claw off, since we had no idea of where we were, and the land was a lee shore, but I was determined to plant my feet on dry soil. Another night, I was convinced, would bring disaster.

All day we sailed, with me clinging desperately to the tiller, but nightfall came when we were still several miles offshore, with no idea of where we were. I'd sailed into the perfect trap. Now I really would have to start clawing out to sea.

Then a lighthouse winked. Its beam pierced my heart with unconditional love. What a wonderful thing humanity was, all that effort and expense, just to reach out to wandering sailors, to shine a light on their darkness, and guide them safely back into society's bosom. I wanted to kiss everybody, starting with Colleen, but instead I got out my stopwatch and timed it.

This show of professionalism bolstered my flagging self-esteem, but I thought wryly of the young couple I'd bought the watch from a couple of months earlier. They had turned back from a passage to New Zealand in their clinker-planked Folkboat after encountering a fierce storm. They'd shown me their bruises, their soggy charts and torn sails. I could sense their distress, and refrained from making critical comments, but privately I thought they were a bit soft.

The light was on Point Stephens, at Fingal Bay, just south of Port Stephens. I knew the port well, for I had been there many times on the millionaire's yacht, but I was uncertain how safe the shallow entrance would be in this huge sea. The decision was soon taken out of my hands, as the wind rapidly backed NE, then NW, increasing in ferocity as it did so. There was no way we could get in now.

I should have just laid ahull, drifting quietly in the lee of the land, but shame at my earlier bad seamanship led me to commit another folly. I ran back out to sea for several hours, the wake ablaze with phosphorescence, looking like a rocket ship flying through space on the dark side of the moon. The sky had cleared, and the stars, vivid in the blackness, added to this effect. Eventually, poor little *Poeme* was overwhelmed once more by wind and waves, and I had to take all sail down.

By dawn, the wind was in the SW and solid. Colleen, an experienced motorcyclist, estimated the wind-speed to be about 45 knots. The waves were very steep. I have no idea how high they were, only that the sight of them was impressive. The sea was covered in long streaks of foam again, and in places, where the sea broke, there were large patches.

The sky was clear and the sea looked brilliant under the glaring sun, though its awesome beauty was only dully registered. Later, I would discover from local trawlermen that the seas were opposing a 4-knot southerly current, which was sucking *Poeme* south against the wind. I had known about this current before I set sail, everybody did, but my exhausted, traumatised mind forgot. I assumed that we were being blown north. *Queensland, here we come*, I thought. It was a cheerful thought, or as cheerful as possible under the circumstances.

*Poeme* was repeatedly knocked down, surfing sideways with the crests, masts almost horizontal, to judge from the angle inside the cabin. We had no intention of going outside to look. We did not keep watch at all. If there were any ships in the vicinity, and there must have been, since we were in a major shipping lane, there was nothing we could do about it; we could not have manoeuvred out of their way, and had no way of communicating with them.

Anyway, we had more important things to worry about. The companionway doors, a poor design, had smashed earlier in the day when I was thrown against them. If *Poeme* capsized now, the boat could flood. I though dully of ways to barricade the hole, but that would mean we were stuck inside, as the large hatch above had been screwed down.

Remarkably little water came in through the open companionway, but plenty poured in through the planks, especially that cracked sheer plank, when waves broke aboard, which they seemed to do with monotonous regularity. Pumping was almost continuous. I then discovered that the cabin top had shifted slightly, cracking the carlin, the longitudinal timber that holds it to the deck. Looking closely, you could see daylight through the joint.

Land had long since been lost to sight. We were alone once more in the watery wilderness. We ceased to wonder when the gale would end. We had been wet and cold for so long, it seemed as if we'd been born that way, creatures of primordial slime. And by this time, our skins *were* covered in slime, and in swarms of angry rashes, plus some boils.

No attempt to control the boat was made. We just pumped and clung to our bunks. Except that Colleen cooked something every day. Once, in an act of pure heroism, she produced curried eggs and buttered potatoes for lunch. She was determined that we would eat something hot every day, and persevered with this self-imposed task without fail, paying a heavy penalty in scrapes, bruises and burns.

She even produced a special treat on 12 April, my 22nd birthday. How I would have coped alone is questionable, as I was in a sort of trance by this time. I may have eaten but I would never have cooked. My strength would have been seriously undermined, and I would have had to do all the pumping. I may not have survived. Colleen, a registered nurse who loved skydiving, was much tougher.

Nightfall brought the added misery of blindness. The kerosene lamps, even those in the cabin, would not stay alight in these conditions, and we had not brought any electric torches. Enveloped in total blackness, the violence continued. The wind was much stronger now, Colleen said about 70 knots, and it screamed like a banshee in the rigging, making the whole boat shake. The sea was almost white with foam, as if we were in a snow-storm.

Suddenly, the mainsail broke out of its lashings and began to flog. It felt as though it might tear the mast right out of the boat. I realised it would quickly shred, taking with it any chance of sailing home, but I still hesitated in the shattered companionway, in the grip of conflicting emotions. Then I did what had to be done, leaping out on deck, but without my safety harness. In my tiredness and fear, I'd been unable to sort it out, and anyway it always got tangled. I hastily bundled the sail up.

The reality of being on deck was less frightening than the prospect had seemed, but I was still relieved to get back into the cabin. Then the mizzen broke loose. This time I went up immediately, tied it very firmly, and re-tied the mainsail as well. I considered doing the jib, but baulked because the foredeck looked too exposed. Anyway, we had a couple of spare headsails.

Colleen then decided that she needed some fresh air. Despite my pleading, she clambered out and clung to the mizzen mast for half an hour. While she was out there, the boat was twice spun completely about by breaking seas, the bows pointing in the direction that the stern previously had.

The severity of this gale was later confirmed by reports from two yachts that capsized while lying ahull nearby. The 36' timber sloop, *Surprise,* a John Illingworth design skippered by an Italian solo sailor, Ambrogio Fogar, had already weathered Cape Horn when floored by the Tasman Sea. *Surprise* suffered damage to the spreaders, but managed to sail into Sydney Harbour under jury-rig. The other vessel, a Duncanson 34 from

Hobart, recorded wind strengths of 70 knots, confirming Colleen's estimate, before losing the mast and much deck gear in a violent roll. *Poeme*, being tender and lightly ballasted, simply lay over and surfed the crests sideways. I am not suggesting this as a survival technique.

Finally, although at the time we were unable to believe it, the wind began to ease. All sail was made over a lumpy sea, westwards without question. Australia was now the Promised Land, the Shangri-la I'd always been looking for. Anywhere in Australia would do, provided we could reach it without drowning. I steered and Colleen pumped. *Poeme* leaked more than ever.

At the end of a long day, a lighthouse came into view. Once again, I timed it. Three flashes every twenty seconds. I got out all my charts north of Port Stephens. It could only be Smoky Cape. Since the chart showed clear approaches, I decided to carry on through the night, but the seas soon became noticeably steeper. Colleen looked at me quizzically.

'The chart shows clear water,' I said, 'so we must be in the grip of a current. It'll ease off when we get closer in.'

Instead, the waves became even more steep. *Poeme* almost fell down their faces. They began to break heavily, and crests the size of small houses rolled past. Then the inevitable happened, and one of these crests landed on *Poeme*. Tons of water swept over the decks, carrying away everything that wasn't double-lashed, and *Poeme* came back up facing out to sea.

I glanced at Colleen. For the first time in this ridiculous adventure, her face did not reflect the imperturbable good humour that had buffered me from the worst of my panic. What I saw there probably reflected the pure terror I felt. Without a word, I heaved in the sheets and we hauled our arses out of there. Now that *Poeme* was facing the waves, they seemed *really* terrifying, and impossible to sail over, but up the little boat went, again and again. It was a magnificent effort, but I was too frightened to be proud of my ship.

Then the clew pulled out of the jib, and *Poeme* fell back broadside to the waves. Driven by necessity, I dragged out a spare jib and scrambled forward. Perched up there on the bottom of the upturned dinghy, with my feet braced on top of the solid guard rail, only the grip of my buttocks between me and the sea, I completed one of the quickest headsail changes in history, hoisted the new jib, and scrambled back to the cockpit. While I was up there on that exposed foredeck, I was aware that this was the raw material of a heroic sea-story, the sort I usually relished, but rather than feeling proud, I was ashamed of my corrosive anxiety, and the mess I'd made of everything.

As we sailed back out to the east, the seas rapidly calmed. 'Some current,' said Colleen, with, I thought, a certain amount of emotion, but being the person that she was, she immediately set to fixing us some dinner.

Since the prevailing wind was holding us off that patch of horrible, vertical water, we hove-to by backing the jib and crawled below to rest. Throughout the night, the wind eased, and by dawn we were becalmed on gently undulating swells. The coast was clearly visible in the distance, tantalising, evocative. Had green hills ever looked so lovely?

There was not a breath of wind. Once more, a little engine would have been worth its weight in gold, or a rig with plenty of sail area and lots of light-weather sails, but at first, we were happy just to lie on the deck, exposing our slimy rashes to the healing rays of the sun. Glorious, glorious sun, silence, stillness. No more howling winds, crashing seas, lurching boat, or soggy bunks. Well, the bunks were still soggy, but the crew were lying on deck.

We were becalmed within sight of land for four days. With increasing frustration, I chased every little breeze. At night, a land breeze would fill the sails and *Poeme* slipped along quietly. By dawn, we would be close to the lighthouse on the top of the perpendicular cliffs, convinced we would be at anchor for a late breakfast. Then the rising sun dispersed the wind, and the current began to suck us south again.

Once we drifted past what appeared to be a snug little harbour, with breakwaters and all, but since there was nothing like that on the chart, I decided it must be an optical illusion. 'A pile of rocks,' I said, 'an accident of perspective.' I was also puzzled by the sight of several fishing boats disappearing shoreward around the southern end of the cliffs, when the chart showed no bay there, just a long, exposed beach. I concluded that they must be fishing *very* close inshore.

On the second night, I began to suffer hallucinations through lack of sleep. They took various forms. A small figure, who looked remarkably like Antoine de Saint-Exupery's *Little Prince,* or maybe another of my favourite characters, *Rupert Bear,* swung merrily in the rigging, wearing checked trousers, a scarf, and a jaunty little cap. That was sort of fun.

Once, I became convinced that we were sailing uphill, and on another occasion that we were off the coast of Japan, which felt weird. More insidious, the squeaking rudder pintles turned into carping voices that bickered for hours. When I could take no more,

I'd shout, 'Shut up.' The sound of my voice usually pulled me back to reality. On one occasion, one of the voices said, '*When he's half asleep like this, it would be easy to stick a knife in him and roll him overboard.*' I woke up quickly that time.

Colleen usually stayed below at night, as she couldn't steer the boat and I had no patience to teach her. What she thought, if she heard my ranting, she didn't say. During the day, with Colleen around, I managed to keep a better grip on reality, but one day I thought I'd succumbed. Colleen was up near the mast, just behind the dinghy, sitting on the toilet bucket, when suddenly a huge manta ray leapt out of the water beside her. To my utter incredulity, Colleen's arms turned inside out, like a bird raising its wings, and she began to squawk. She hadn't told me about her predisposition to dislocating shoulders. Despite my panic, she instructed me how to relocate them, and then, unperturbed, continued her toilet. I retired to the cockpit, badly shaken. Thank goodness it hadn't happened during the storm.

Not long after Colleen's bird impersonation, a military plane flew low over the mast-head. It was so low that we could see the faces of two men peering at us from its cockpit. We waved. It circled around again and came in even lower, almost at masthead height. The noise was deafening. One of the men held up a microphone, pointing at it. *Ah...sorry, buster, me sea gypsy, no radio.* I held up my hands, palms outwards, the international gesture of apology.

On the next pass, the pilot pointed out to sea. There was a little boat out there, not dissimilar to *Poeme*. I frowned. *What does he want me to do, tow it? Would I be bobbing here like a moron if I had a motor?* Then another plane flew past and dropped a bomb near the other boat. A geyser of water shot up into the air. *Shite.* We were in the middle of an active bombing range.

We later heard that the dates of bombing exercises were posted in the Admiralty Notice to Mariners, and broadcast over VHF radio, but we had access to neither. I had never heard of such a thing, but discovered later that such exercises are common practice up and down the Australian coast. *Our* plane continued to make passes overhead, shepherding us until we drifted out of range.

On the fourth day, a moderate southerly finally arrived. *Poeme* raced north all day, past the cute little place that had the illusion of being a harbour, past the place where all the fishing boats mysteriously disappeared shoreward, past, finally, those distinctive, perpendicular cliffs. This time we were going to get in, even though it was dark again.

I took a bearing on the light with my hand-bearing compass, estimated my distance off, marked my position on the chart, and drew a course that would take *Poeme* straight into the bay. I was astounded to discover, when I tried to steer this course, that *Poeme* was pointing at the cliffs. Swearing, I turned the boat away and returned to my chart, but came up with the same result.

'Something's happened to the main compass,' I said, 'we'll have to steer using the hand-bearing compass instead.' But that produced the same results. Increasingly frantic, and swearing profusely, I wrestled with chart, kerosene lamp, which kept blowing out, and pencil in the cockpit, while trying to steer at the same time. Colleen, unperturbed as usual, handed up coffee and dessert. Poor little *Poeme* went around in circles.

The result was always the same. There was only one rational explanation, both compasses had gone bananas. I decided to sail in by eye, guessing my distance off the cliffs. It was scary, but soon we were in the bay and breathing easy, safe at last. Tender thoughts about the little ketch flooded my mind. I would do all the work that should have been done before setting out, turn *Poeme* into a real, tough little ocean voyager. I even thought of colour schemes, and a beautiful, new, hand-carved, varnished tiller. My hand caressed the wonky old tiller.

But where was the short breakwater mentioned in the Pilot Book? And the bay seemed a lot smaller than expected. I was just about to put the anchor down and wait for daylight, when I discerned the faint outline of some boats moored ahead. *Good,* I thought, *I'll go over and anchor near them. They must be in the shelter of the breakwater, even if I can't see it.*

Just before reaching the group of boats, jogging along under jib and mizzen so as to make a gracious arrival, *Poeme* struck the rocks. Instantly, everything was transformed. Elation evaporated as the hull pounded and staggered with each passing swell. The sound of splintering wood made me weep, but I mustered my last strength to try and save the boat.

All my theoretical knowledge, including the way I had used it prior to departure in preparing warps, knives, etc, became vital instruments in our battle to save the boat, perhaps even to survive, for although we were close to shore, Colleen's dislocating shoulders and my hallucinatory tiredness made us poor candidates for a long swim.

Over the side with the dinghy, in with the oars, anchor and warp. Somehow, I got it together without swamping, and rowed off into the night, dropping the anchor when all the scope had been paid out. Back alongside, trying to get a second anchor prepared, the dinghy suddenly disappeared beneath me and I found myself swimming in the cold sea, with full wet-weather gear and boots on. I could barely stay afloat. It took all of our combined strength to get me back aboard *Poeme*.

Every time the swell lifted the boat, the two of us heaved on the warp, but the keel continued to pound on the rocks. Mouths dry with fear, we re-hoisted the mainsail, to heel the boat and reduce draft. More warp came in, much more, but still the rocks clutched at the keel. We later discovered we were hauling poor *Poeme* further up the rocks. In my exhaustion, I'd rowed the anchor out in the wrong direction.

Defeated, we lay on the deck, contemplating what to do next. Dawn would come soon, and with it, perhaps, the chance of rescue, if the boat didn't break up first and throw us into the sea. 'Perhaps the tide is coming in and we can float off,' suggested Colleen.

'Don't be stupid,' I replied. 'Tides only happen in rivers; this is the ocean!' I wondered why she gave me such a peculiar look. I had lost the plot in more ways than one. Then a diesel engine roared into life nearby, navigation lights came on, and a boat began to move slowly across our bows. We shouted, waved our lantern, became slightly hysterical, but the trawler chugged past without stopping. The sense of being abandoned was overwhelming.

A great howl rose up inside me. So near to the end of our dreadful, isolated nightmare, I felt that to endure another moment was unbearable. Stumbling aft along the jerking deck, I crawled into the cabin, where water now surged over the bunks. Once again, my earlier efficiency paid off, for I quickly emerged holding two red flares. The first flooded the night with its eerie glow. Although desperate, I still felt embarrassed and melodramatic, standing there, so close to the beach, which glowed in the moonlight, holding the flare aloft. My heroes didn't do things like this.

The trawler never altered course. Swearing, I fired the second flare. They were the type that is activated by removing a cap from the bottom and striking it, flint-like, across the top. This time, however, I neglected to remove my hand. I watched the flare explode against my flesh in dull amazement, until I belatedly realised it was my responsibility to do something about it.

I was still swearing at the trawler, the weather, the rocks, the pain in my hand, at the whole bloody, puzzling mess, when a small rowing boat appeared, coming from the opposite direction to where the moored boats were. There were two middle-aged men in it, and I remember looking down at their weather-beaten faces and rumpled clothes, wondering where the hell they had come from. I later realised that they had simply rowed out from the beach. It was such a calm and peaceful night that there was no surf in the bay. The long white beach shimmered like a ghost in the distance. They were local fishermen, it transpired, and one of them quickly took charge of the situation.

'What's the matter, matey?' he said. His partner watched us silently, his sardonic gaze taking in the details of the battered, rust-stained ketch, and its dishevelled crew. I thought the problem was pretty obvious. Later, I realised that it must have seemed incongruous to them. How could anyone get into trouble on a clear, moonlit night like this, in a bay wide open for navigation except for one little reef, marked clearly on the chart?

The tide *was* rising. After laughing at the position of our anchor, the men cut the rope and towed *Poeme* off, putting us on a nearby mooring. I pumped out the water in the bilges, and they stayed reasonably dry, though I soon discovered that I had to pump regularly to keep it that way. Before we left Sydney, the bilges had stayed dry when *Poeme* was lying on its mooring.

'Where are you bound for, matey?' our saviour asked.

'Well,' said the Great Navigator, 'we left Sydney for New Zealand twelve days ago, but were driven back by a storm. Now that we're at Smoky Cape and halfway to Queensland, we might just go there for the winter.'

'This ain't Smoky Cape, mate,' he said. 'This is a place called Currarong, 100 miles *south* of Sydney. That's Cape Perpendicular out there. You're hundreds of miles off course.'

That little pile of rocks we'd seen further south was a port called Ulladulla. Jervis Bay, almost as big as Sydney Harbour, lay just around the corner, at the southern end of the cliffs. *That's* where all those fishing boats had been heading.

'But, but...' I stuttered, realising suddenly why none of my bearings had worked out, yet unable to understand the error. 'But I timed the lighthouse, three flashes every twenty seconds.'

'Yep, that's Point Perpendicular Lighthouse. You're lucky you missed all the reefs out there if you weren't going by the chart. There are rocks scattered everywhere out round the point. Not to mention the Sir John Young Banks just offshore. Only twenty-six feet of water on them, they're treacherous in heavy weather. Boats have been known to fall down the face of waves and pitch-pole in the troughs on those banks.'

He paused as Colleen and I looked at each other in sudden comprehension. Then he added, 'Well, I reckon you've had enough by the look of you. Better come ashore with me for some breakfast and a shower. We'll look for your dinghy later, it's probably on the beach somewhere.'

We were unable to walk straight when we got ashore, staggering all over the beach like drunks. It was daylight by then, and our saviour could see how young we were, and how badly beaten up. Our skins, though improved, were still a mess. Our arms and legs were covered in bruises and rashes, and we both had boils on our torsos.

'You kids have really taken a hiding,' he said, and the concern in his voice brought tears to my eyes. He cooked us up a huge breakfast, and later took us down to the beach to look for our dinghy. 'Anything you kids want,' he said, 'just ask.'

We found the dinghy and oars, and then rowed back out to *Poeme.* The bilges needed pumping again. Sitting on the cabin roof in the sunshine, we conducted a post-mortem. It had never occurred to me that there could be *two* lighthouses with the same sequence, even if one was well north of Sydney and one south. There were marked variations in length of flash, and gap between them, but the variations were too subtle for me to notice.

I'd also forgotten about the East Coast Current. While *Poeme* had skidded sideways down the face of waves, apparently being blown north, that insidious current had dragged us south at a rate the old boat would have been proud of. No wonder those waves had been steep. Our storm, we discovered, was an East Coast Low that had collided with a deep, ex-tropical cyclonic depression coming down from the Coral Sea, causing prolonged, severe gales.

Down below, we examined the damage to the cabin structure and our stores. Most of the tins, so carefully varnished, just as I'd seen done in the *Sandefjord* movie, and read about in books, were dented and rusting. All of the 'easy-opening' cans of fish had opened, their contents mostly pumped overboard, though I was to hunt stray odours for some time. Some of the flat sardine cans had even managed to slip between the planking and the fore-and-aft stringers when the boat had flexed, and were now irretrievably wedged. Both of my cameras were ruined, and the few photos I'd taken of the trip lost.

'So, what now?' asked Colleen, forever game.

'I don't know,' I said, looking down. 'I might just stay here.' My voice carried as much of an apology as my fragile ego could offer.

'Well, I'll leave then,' she said, and within half an hour she was gone. I never saw her again.

I really *didn't* know what to do. I sat in the cabin for several days after Colleen left, just thinking. The *Otaha* fiasco had shaken me briefly, but this time I was really rattled. I had created a beautiful story for myself, in which I took my place alongside the likes of John Guzzwell, Robin Lee Graham, John Sowden and Julio Villar, sailing their simple little boats across oceans. Somehow, I had lost the magic in Sydney, but I still had the belief that it was *out there* somewhere. All I had to do was sail across an ocean like my heroes. It was shattering to realise that I did not have their courage or ability.

# Chapter Eleven

# Return to the Sea

For a few months, I had no sense of direction, but eventually decided I would have to go back to sea. As Susan Hiscock said, after Eric died aboard *Wanderer V* in 1986, the solution was to sail. I might be pretty useless at it, but voyaging under sail was my *raison d'etre.* Anyway, I was too old for ballet school.

To add to my humiliation, I learned that a young Japanese sailor, Yoh Aoki, had just arrived in Sydney on his 21' plywood ketch, *Ahodori*, after a voyage from Japan via San Francisco, Cape Horn, Buenos Aires, Cape Town and the Southern Ocean. I later met him at the Cruising Yacht Club in Rushcutters Bay, and realised that *Ahodori* was a lot stouter that poor, tired, old *Poeme.*

That is not saying much, and Yoh's voyage remains one of the most audacious in history. He suffered several structural failures, including his keel, which began to wobble alarmingly after several knockdowns approaching Cape Horn. The boat had to be substantially rebuilt in Argentina. His book, *Yoh and Ahodori*, has recently been translated into English.

I wasn't going to do it on *Poeme,* however. I had no desire to drown like some Viking, going down with sword in hand and an oath upon my lips. The boat needed a rebuild, and some sort of engine, given its modest ketch rig, or maybe set up as a gaff-rigged cutter with a bowsprit, its original configuration. I'd learned that to sail engineless you need a boat with good performance, plenty of sail area, a willingness to change sails as soon as needed, excellent seamanship, *and* a smidgen of luck. That would take a lot of work and money. Given that *Poeme* was old, small and cramped, it didn't make sense.

*Yoh Aoki and his 21' plywood ketch, Ahodori, sailing in Table Bay off Cape Town. Photo: courtesy Yoh Aoki.*

First, I had to do something with *Poeme*. I had the boat towed around to Currambene Creek, inside Jervis Bay, where there was a township called Huskisson, an old timber ship-building settlement. The last of the old-school boat-builders there, Alf Settree, who still built fishing boats with grown timbers, carefully selected in the forest for their shape, offered me a free mooring.

'You look like you need a break, son,' he said. Alf was one of those old-fashioned Australians, an honest, hard-working, modest and generous man, who wore checked flannel shirts and grey working shorts every day of the week, except for weddings, funerals, christenings and Christmas dinner. Not one to extol his own virtues, I heard from others that he was an exceptionally-talented boatbuilder, revered among the fishing community.

I lay alongside his boatshed for a day, cleaning up the mess and washing everything in fresh water. Amazingly, I found a dead rat in the lazarette. Whether it had drowned in the storm, or died from indigestion after nibbling a lifejacket stowed there, I'll never know. How we co-existed on that small boat without meeting each other is another unsolved mystery.

At one stage, a shadow fell across me. I looked up to see a broad-chested, black-bearded, naval officer, resplendent in uniform, gold braid everywhere. Being not that long out of South Africa, anyone in uniform scared me silly. I immediately thought of the incident on the bombing range and began stuttering apologies.

'I'm navy, not bloody air-force, let alone the army, or whoever the hell it was,' said the officer, laughing. 'We don't care if you get yourself blown up. I'm Max Kean. I remember this little boat from Elizabeth Bay in Sydney Harbour. Never seen it off its mooring before.'

Then he laughed. Max was 36, a passionate sailor, and he and his partner, Lindy Edmunds, lived in a small cottage at Hyams Beach, at the southern end of the bay. He took me home for a hot shower, where Lindy did my laundry and fed me a fat, juicy steak. That night I slept in a warm, dry bed for the first time since whenever. When Max dropped me back to *Poeme* in the morning, I was purring like a cat.

I spent several months living aboard *Poeme* in Huskisson, contemplating the future. When the wind howled and rain splattered on the deck, making *Poeme* tug at the mooring, I lay on my bunk drenched in sweat, unable to sleep. Depression began to cloud my mind. Alf was endlessly generous, but it was Max and Lindy to whom I turned most, for there was a warmth and gaiety in their cottage that was infectious. There were times when I realised that I was being a pest, and tried to stay away, but my resolve never lasted long.

Eventually, I borrowed some money and trucked *Poeme* to Sydney, painted the hull bright red with black trim, just like Bernard Moitessier's *Joshua*, and listed the boat for sale with a broker. Moitessier claimed that red was a strong colour, the colour of desire, and the new colour scheme was very effective; it worked so well that I almost kept *Poeme* for myself, but two young dreamers came along and snatched it within days. I never met the new owners and have not seen the boat since, but I often wonder what became of it, and hope it didn't drown the poor boys.

'Come back to Sydney and build yourself a decent boat,' said the ever cheerful and optimistic Gavin, who'd started building a ferro-cement yacht in Terry Hills, out near Duffy's Forest, since the time he'd helped me work on the aborted catamaran project. I thought that might be my best option.

I had a clear idea what I wanted. There's nothing like first-hand experience to sharpen one's vision. It would be steel-hulled, like *Joshua*, though smaller, about 28', with standing headroom under a short coach roof aft of the mast, housing the galley to port, and chart table plus toilet to starboard. The latter would also serve as a locker for wet-weather clothing when at sea. There would be a flush deck forward of the mast, something like Clive Rouse's *Clipper*, or a slightly larger version of John Sowden's *Tarmin*.

My boat would have a single junk sail on an unstayed mast, like *Jester.* No more clambering around the foredeck in a storm, or worrying about all those rigging swages, clevis pins, split pins, and all the other bits and pieces that hold a bermudan rig up. The dinghy would stow on the coach roof aft of the mast, where dinghies belong at sea.

There would be a small, hand-cranked, diesel engine, and electric lights. To hell with smoky, smelly oil lamps that added injury to insult by blowing out just when you needed them. They are OK for an alternative source of light in the cabin, I decided, to reduce electrical consumption, or emergency back-up navigation lights.

The rudder would be mounted on the transom, easy to access and maintain, tiller-steered of course, with a trim tab on the trailing edge, linked via a differential linkage to a vertical-axis windvane. Above all, the boat would be massively strong. I realised that I was not a cavalier adventurer like Julio Villar.

It was a good plan, and should have been straightforward, but I was talked out of it by Gavin and other well-meaning friends, who persuaded me to build in ferro-cement like they were, in the same boatyard, and adopt a standard bermudan rig.

Steel rusts, they said firmly, and junk-rigged boats sail like dogs. In rebuttal of my arguments for junk rig, they extolled the virtues of roller-reefing jibs and halyards led aft to the cockpit. I acquiesced, but it was a mistake. I have always been too easily influenced by the people I like. I bought the plans of a 30' yacht and construction began, but my heart was never in it.

My depression deepened as the weeks went by. I thought about light-hearted Julio Villar, who just ignored *Mistral's* structural problems, blithely setting off on passages that made other sailors quail, and of Tom Corkill on his fragile little trimaran, saying, *the bigger the risk, the greater the adventure.* Then, after *Clipper* capsized, he returned to Durban on the equally-dubious *Ninetails.* Even John Guzzwell, that consummate boatbuilder and seaman, admitted it wouldn't take much to knock a hole in *Trekka's* 5/8" red-cedar hull. Guzzwell hardly kept a lookout at sea, and often sailed unlit at night. Approaching Panama, he began to see a lot of ships. Although he felt it unlikely a ship would run *Trekka* down, given all that ocean, he casually hung a kerosene hurricane lamp in the rigging before going back to sleep. I knew how effective *that* was. Were they all mad? I was beginning to doubt my heroes.

*Standing dubiously inside my new ferro-cement hull in Terry Hills. The tin shed I lived in was behind the back wall of this shed. Photo: Gavin Fox.*

I was living in a tin shack without insulation that I'd knocked up behind my new boat, and when the wind blew, it moaned around the eaves and through the surrounding trees, while small branches fell onto the roof with a resounding crash. I lay awake for hours at night, sweating with fear, having flashbacks to those desperate days on *Pocme* and wrestling with my doubts. I was deep in the grip of post-traumatic stress disorder, but had never heard of it. I just felt like a coward.

When Cyclone Tracy devastated Darwin on Christmas Eve 1974, the company I worked for, Stewart and Lloyds, which had a branch in Darwin, asked for young, single, volunteers to go up and assist. Many of their regular staff, having lost their homes, had gone south with their families. I volunteered immediately, planning never to return. I would build a small steel boat in Darwin, as originally intended, and head west across the Indian Ocean.

It felt liberating to be so far away from Sydney, and at first, I revelled in the move. However, six months later I was on my way back, unnerved by the chaos in Darwin. A lot of people were suffering from post-traumatic stress after the cyclone, and the city was a madhouse. The chaos had its entertaining moments, and I could dine out on stories about some of the antics that went on, but underneath my humour lay something close to hysteria. The widespread destruction of buildings added to my difficulties. It was like an externalisation of how I felt.

At the Botanical Gardens in Brisbane, I found a vibrant community of international voyagers moored fore-and-aft on the pile berths. The atmosphere reminded me a lot of Durban. I had the feeling that if I could just get a decent boat underneath me and come to the Botanical Gardens, I'd be OK.

That Botanical Gardens community is long gone. Voyaging sailors tend to be older and more affluent these days, and sail larger, more expensive boats. They prefer berthing in marinas with walk-on, walk-off, floating pontoons, a number of which have been built around Moreton Bay. The pile berths were taken over by local boats that couldn't afford marina fees, but recently the state government built a low bridge across the river downstream of the berths, blocking access for vessels with masts, and began removing the pile berths.

I also found Tom Corkill up nearby Tingalpa Creek, refitting *Ninetails* after his 5-year circumnavigation. Married now, he had a little more flesh on his bones, but was otherwise unchanged. I was shocked, though, to see a 25hp outboard motor bolted to *Ninetails*' aft beam, as Tom had always been passionate about sailing engineless, keeping everything as simple as possible. 'Anchorages and harbours are too crowded these days to sail engineless,' he said. 'The best days of cruising are over.' He was thinking of building a house and starting a business, selling *Ninetails* and getting a trailer-sailer.

Best of all, I spent three weeks with Adrienne at her property just south of Brisbane. She was as vibrant, irreverent and thrilling as always. We demolished numerous bottles of wine, and she regaled me with more stories of her years on *Kelasa*. For a short period, I felt reconnected with the magic.

Then, reluctantly, I returned to Sydney. I wasn't sure what I was looking for, though I knew it wasn't the ferro-cement hull that languished in its shed. Perhaps a *successful* voyage would provide the answer, but I had no idea how to achieve it.

A dreary year followed, and I began drinking too much alcohol to dull the pain. I'd sworn I wouldn't drink alcohol, due to memories of my father's behaviour when drunk, but I'd long discovered its short-term anaesthetic properties for anxiety, and its value as a social lubricant. But alcohol is a dangerous drug for people like myself, just as it had been for my father. It unravels you. I was never aggressive (perhaps because I didn't have an impossible teenager to provoke me), but became somewhat self-destructive, crashing motorbikes, falling off fences, etc. Eventually, I fell off a moving train. My neck still hurts, but the incident ended my drinking days.

At the suggestion of a well-meaning friend, I saw a doctor, which turned out to be a bad decision. He prescribed anti-depressants, which did nothing for my sadness but an excellent job of damaging my physical health. As well as being sad, I became dull, lethargic, overweight, and developed gut issues that still haunt me. I also suspect that one of the drugs was responsible for another, life-threatening, illness that developed later.

Then, in mid-1976, an opportunity to go on another ocean voyage unexpectedly presented itself. After Barry Lewis brought *Ice Bird* back to Sydney in 1974, he'd gone off to skipper a cargo barge in Rabaul. David was in Hawaii, involved in the *Hokule'a* voyaging canoe project, and *Ice Bird* was languishing on a mooring in Woodford Bay, up the Lane Cove River. Barry's girlfriend, Ros Dawes, myself and others, had been sailing *Ice Bird* around Sydney Harbour and maintaining the boat.

With David's blessing, Ros came up with the idea of sailing *Ice Bird* to Rabaul, to join Barry, and asked me to come. I never hesitated. She also asked Gordon Lewins, another member of our gang. We were all 24 years old.

Although very different, the three of us made a good team. Gordon was pragmatic, calm, and good on the tools, while Ros's razor-sharp intelligence had an answer for every fear I could imagine, and she had a bloody-minded determination to succeed that would accept no excuses. As for me, I had a good theoretical understanding of voyaging and a modicum of experience. Ros was the skipper and I agreed to be the navigator. (What, I hear you say, WHAT? Make sure you don't end up in Jervis Bay.) I didn't tell Ros that I didn't actually know how to navigate. But I had a plan.

I took *Ice Bird's* venerable sextant down to the beachfront at Manly, on the harbour's north side, early one morning, and took a series of sun sights while perched on the back of a public bench, ignoring the glances of curious onlookers. Using Eric Hiscock's book, *Voyaging Under Sail*, which contains the clearest instructions for celestial navigation ever written (the essentials are covered in a mere four pages), and the *Sight Reduction Tables for Air Navigation*, I worked the sights out, overjoyed to obtain a series of position lines that crossed Manly Beach. *Eureka!* Why I didn't do this in 1974 I have no idea, though it would have helped if I had not sold the sextant I'd been gifted. I never did work out how to use that box sextant with its bubble horizon.

*Ros Dawes in 1976. She is now married to Barry Lewis and they have sailed many miles together.*

I couldn't take a noon sight for latitude from my park bench, as by then the late winter sun had already crossed the land, but I knew they were easy. All you have to do to get a fix, giving latitude and longitude, is advance your morning position line, using a parallel ruler, the distance and direction you estimate the boat has travelled between taking the morning sight and local noon. Where that position line crosses the latitude obtained from the noon sight, is your current position, give or take a few miles, due to variables such as current and drift. The noon fix, as it is known.

I also learned from Hiscock's book how to use Baker's position line charts, and bought myself a supply from Boat Books Australia, a wonderful shop that sells sailing books and navigation supplies. Over the years, I've spent a fortune in that place. Once I'd mastered basic celestial navigation, I felt like Christopher Columbus, or perhaps Captain Cook. Columbus was a somewhat rudimentary navigator; when he arrived in the West Indies, he thought he'd reached India. *Been there, done that.*

The next step was to prepare the boat. *Ice Bird's* galvanised rigging was rusty, so Ros and I replaced it, wire by wire, without taking the mast down. We were helped by the fact that Barry had left aboard a list detailing the lengths of each wire, allowing us to get them made up before starting the job. Given my fear of heights, it was an ordeal I was relieved to complete. To complicate things, there were always speedboats buzzing past. One in particular, an unmistakable bright-red beast, seemed to appear almost every time I went up the mast. I was convinced that the driver had a waterfront property nearby, and jumped in his boat when he saw me ascend. I am *not* paranoid.

We also fitted a new, mechanical speed and distance log, and I placed the readout dial on the flight between two of the companionway steps. I thought it looked neat, and it gave a short, straight run for the cable between impeller and readout dial, which is desirable.

'That's not a good place,' said Gordon. 'Somebody may kick it.'

'What sort of dickhead would do that?' I replied. I kicked it myself on our first night at sea.

Since we were heading for the tropics, we removed the steel plates that David and I had bolted over the portholes in 1972. In our rush to get away, we never got around to plugging the bolt holes. 'We can do that as we go along,' I said confidently.

The last major job was to take *Ice Bird* out of the water and paint the bottom with anti-fouling paint. Or so we thought. One week later, we had to slip the boat again in a hurry, when a through-hull pipe, just above the waterline fortunately, crumbled to rust in my hands. Colin Putt, a friend of David's, a respected scientist, mountaineer, and former crew of the legendary Bill Tilman, rushed down and replaced all the through-hull fittings.

Many years later, by now in his 60s, Colin put all of his considerable skills to the test by skippering the 65' steel schooner, *Dick Smith Explorer*, to the Antarctic, surviving a capsize on the return passage. After repairing *DSE* in Lyttleton, New Zealand, Colin sailed the ship back to Sydney with just his wife, Jane, as crew. As Tilman once said, Colin was a good man to have aboard in a fix.

It was time to go. In fact, it was already quite late, because cyclone season would soon start in the Coral Sea, but we were unable to leave earlier as Ros had work commitments. Rabaul was 2500 miles away, and it was anyone's guess how long it would take to get there. Departure day was set for Saturday, 11 September 1976. The day dawned cold and windy, with a blustery 25-knot westerly whipping the trees outside my bedroom window. It was exactly the sort of wintery day that I hated so much. Spring comes late to Sydney.

Memories of my first, bleak days in Sydney were never far from my mind, and I wanted to burrow under the blankets and stay there, but Ros dragged them off. 'Come on,' she said with a laugh, 'destiny awaits.'

A collection of friends waited too, including a sun-tanned David Lewis, freshly returned from sailing the Polynesian double-canoe, *Hokule'a*, from Hawaii to Tahiti. He had aged shockingly after his Antarctic voyage. He'd been a fit, muscular 56-year-old when he set sail in 1972, easily pretending to various conquests that he was 10 years younger, but now he was stout, stiff, and ungainly, largely caused by a bad hip, the result of breaking it in a skiing accident after his return from Antarctica. Prior to breaking his hip, he'd been vigorously active, either sailing, trekking or rock climbing. He'd been an active mountaineer in his youth.

But he seemed happy and healthy, with his deep, tropical tan, and was accompanied by another beautiful girlfriend, Yvonne Liechti, from Canada. They had recently purchased a house on Dangar Island, on the Hawkesbury River north of Sydney, and I promised to visit them on my return.

David advised us to stick close to the coast initially, taking advantage of the strong offshore wind to drive us north in the lee of the land. It was excellent advice, but I had my own ideas, based not on experience but on a romantic notion. I wanted to get well offshore, where true sailors belonged, out in the deep blue sea.

After motoring clear of the public wharf in Woodford Bay, we removed most of our well-wisher's streamers from the rigging and hoisted sail. Escorted by several curious boats, (*Ice Bird*, in those days, always drew attention) we roared down the harbour, barrelling before the fresh wind. *Ice Bird*, originally dark blue, had been painted yellow in Antarctica, and with remnants of the streamers flying, several large dents in the hull, legacy of the ice, and newly-painted beige decks, the boat must have looked like a galloping camel.

*Departing from the public wharf in Woodford Bay, Lane Cove River, Sydney Harbour, 11 September 1976. I am at the helm while Ros Dawes stands to the right and Gordon Lewins to the left.*

I might add that *Ice Bird* was a pretty basic yacht. It had a tiny petrol engine, hardly capable of moving the boat in a calm, and an even smaller fuel tank, no electricity, and needless to say, no navigation lights, let alone a fridge. It was just a rusty, dented, steel hull, with a crude plywood interior and a simple sloop rig. I loved it.

We shot out between the cliffs on North and South Heads and plunged into a turbulent sea. It was cold and wet on deck, and the seas quickly built up as we headed offshore. *Ice Bird* rolled and yawed, flinging spray across the cabin. Occasionally, a crest broke aboard, and the unplugged bolt holes in the cabin side spouted water into the interior, like those statues of naked, peeing boys once so prevalent before political correctness conquered the world. The sandstone cliffs of Sydney Heads rapidly sank into the bleak ocean.

I was on watch, huddled inside the cockpit coaming, while Ros and Gordon snoozed below. The Aries self-steering gear was struggling to hold the gallivanting boat on course, and all confidence in my newly-acquired navigation skills seemed to have deserted me. We would never survive. If the boat didn't pitch-pole, sink, get rammed by a ship, or catch fire one day when we lit the stove, we would surely crash onto a coral reef and bleach our bones under the tropical sun. My mind had short-circuited back to the day Colleen and I departed Sydney on *Poeme*.

*Departing Sydney, with the Heads in the background. I was feeling as bleak as the weather.*

Then *Ice Bird* gybed violently. Ros and Gordon appeared on deck like jack-in-the-boxes, and we dropped the mainsail, continuing to career along under jib alone. Looking back, it is easy to see we were seriously over-canvassed, and I had been driving the boat too hard.

Later that night, all three of us were on deck again in similarly dramatic circumstances, when the new aluminium pole, holding out the jib at the bow, snapped in two. The jagged end attached to the sail was flying around like a Berserker's axe, discouraging us from going forward to drop the jib, but by sheeting the sail in tight and cowering behind the mast, from where we extended reluctant hands, the halyard was released and the sail dragged down onto the deck. The pole had broken at the point where a saddle was riveted on, to take the topping lift.

*Ice Bird* continued sedately downwind under bare poles, with the Aries self-steering gear holding us on course, and I went below and put my head under my pillow. By morning, we were approximately 75 miles offshore, based on dead-reckoning. The interior of the boat was saturated, and the sea and sky were sludgy grey.

'I want to go back,' I said.

'We'll go into Newcastle and see what we can sort out,' said the cunning Ros, knowing that, if we went back to Sydney, it would be the last she saw of her navigator. Furthermore, Gordon lived in Newcastle, as did another useful friend, Steve Ramsay. Between them, Ros reckoned they could sort out the boat and help her sort out her options.

The wind eased throughout the day, giving us an easy reach into port. It would have been perfect weather for making tracks, if the navigator hadn't been such a damp squib. Mind you, the cabin was pretty wet too.

In Newcastle, the boat was dried out, bolt holes plugged, jib pole repaired, and numerous other jobs attended to. Having access to Gordon and Steve's tools and workshop was invaluable. Ros persuaded me to stay with them for the week, while they were sorting things out, and to put her through a crash course in navigation.

By the end of the week, I'd decided to continue. Going back to Sydney, defeated, was unbearable. Our revised plan was to sail up the coast as far as Fraser Island and then branch out across the Coral Sea. Ros would do the coastal navigation and I would practice celestial, comparing results. Hopefully, by the time we reached the jumping-off point at Fraser Island, we'd be in the groove.

A sharp squall descended on us a few hours out of Newcastle, requiring a quick sail change. I looked so miserable while we were doing this that Ros finally said, 'For God's sake, go below and have a sleep,' which I gratefully did, putting my head under my pillow again and blanking out reality.

When I woke, the wind had eased, and the next morning we were becalmed 12 miles offshore from Port Stephens in bright sunshine. The sea was an immense blue carpet on which the boat snoozed, while we lounged on deck. It was just the intermezzo I needed, and from then on, my confidence began to improve. There were no more thoughts of quitting. Anyway, Ros wasn't going to stop.

The wind returned that night from the SE, a warm, friendly, consistent breeze blowing over our starboard quarter, just the right strength to hurry us north. For the next 17 days it blew unceasingly, never dropping below 15 knots or exceeding 25. We zigzagged up the coast, with the mainsail vanged out to port and the jib often poled out to windward, gaining confidence as lighthouses appeared where they were supposed to, and as my celestial and Ros's terrestrial navigation continued to complement each other. I plotted my positions on Baker's position line charts, while Ros used the standard Australian Admiralty charts. Our calculated positions were never more than five miles apart.

*Gordon enjoying some fresh air on deck. Note the jib poled out behind him, before we broke the pole a second time.*

The only serious incident during this leg of the voyage was a brush with a ship one night. It occurred during the change of watch, a classic time for such incidents, and was exacerbated by a misunderstanding among us as to whether a constant bearing meant we were on a collision course, or whether it meant we were sailing parallel. We all ran up on deck when the misunderstanding became apparent, to see the ship steaming across our bows. We were so close you could smell burned diesel from its stack, and I swore I glimpsed people through the portholes, but perhaps I was imagining that. We had no navigation lights.

We also broke the jib pole again. The first time it had broken without cause, suggesting a flaw in the aluminium, or a stress crack where the saddle had been riveted on, but this time it was my fault. I'd poled the jib out on my watch but failed to go up to the bows and attach a preventer to hold the pole forward. The boat was rolling and yawing severely, and the thought of crawling along the heaving deck, alone in the dark, was too much for me.

*Ice Bird driving hard towards the Coral Sea.*

I should have called Ros or Gordon but my pride was too strong. Instead, I hung on and hoped, with the end-result that the boat yawed just that bit too far, the jib was caught aback, and the pole bent around the rigging like cooked spaghetti. We tried an old timber pole for a while after that, but when it began to crumble, we sheeted the jib flat, and drove on with the squared-off mainsail, just like Joshua Slocum had done on *Spray*. It worked, too, but unlike Slocum's ship, *Ice Bird* needed the Aries self-steering gear to stay on course.

After passing Fraser Island, we took a deep breath and headed for the gap between Frederick and Saumarez Reefs, through which we would pass into the Coral Sea. The next evening, the loom of Sandy Cape Lighthouse, at the northern tip of Fraser Island, was clearly visible, bouncing off low clouds from an amazing 60 miles distant. It gave us comfort that we were approximately where we were supposed to be. There was no more land for Ros to plot fixes on, so we were now reliant solely on my celestial calculations.

# Chapter Twelve

# Rolling Down the Trades

At least the wind continued to blow steadily from the SE, and *Ice Bird* was making excellent progress. Less than 10 days out of Sydney, we crossed the Tropic of Capricorn, and were officially in the fabled SE tradewinds that I had dreamed of for so many years.

On the morning of 30 September, we were just south of Frederick Reef, lining *Ice Bird* up for my first serious navigation challenge. As if to tease us and prolong our agony, the wind eased, but a sloppy sea remained to throw any wind there was out of the sails. I was on watch at about 1000, keeping a leery eye out for the reef, feeling seasick and self-pitying. Then, to my horror, I saw a several large, black objects dead ahead. 'Rocks,' I screamed, jumping to my feet.

But it wasn't rocks, it was whales. 10 Grey Whales, mostly about 30' long, I estimate, were swimming south. Four of them were swimming abreast, flanks almost touching, snouts in a perfect line. They looked majestic, a suave blend of power and grace. All except a little one, about half the size of the others, which was straggling along behind, flapping its tail madly, trying to keep up. The sight of it plucked my heart strings. 'Wait for your baby,' I shouted. This incident lifted me, temporarily at least, out of my depression. This is why you go ocean sailing, I thought. All the fear and discomfort mean nothing when you have an experience like this, which simply cannot be replicated in any other way.

The SE wind picked up again that afternoon, and during the night we passed Frederick Reef. Although it had a 15-mile light on it, we sighted nothing. We should have seen it, and it was an anxious time. Staring into the darkness, wind whistling in the rigging, bow wave roaring, wave crests tumbling, it was hard to imagine that we would see or hear the reef before hitting it. I could only trust that my calculations were correct, and that the analog compass had not developed any new deviation. Today, GPS units have taken the terror out of this sort of situation; as well, I suppose, as the thrill. On that night, I would have given almost anything for such a tool.

*The morning after we passed Frederick Reef without seeing its light. Note the reefed mainsail, with several turns around the boom.*

A morning sight on 2 October placed us clear of danger, but when I advanced it along our dead reckoning course to cross a latitude sight at noon, I received a nasty shock. We were about 50 miles further north than I thought. This meant we should have altered course several hours earlier, and subsequently had to drive *Ice Bird* to windward all day to get back on course. To leeward lay the extensive Swain Reefs, into which it would be fatal to blunder. We were never in imminent danger, but it shook my confidence.

A post-mortem revealed that the sumlog was malfunctioning. Because I had placed the recording mechanism so low, and because I'd cracked the face of the readout when I kicked it, losing the air seal, seawater had seeped in. Gordon had mentioned several times during the night that he thought we were sailing faster than the log was recording, but I chose to rely on the instrument rather than our senses. It was a hard-learned lesson.

Gordon took the log apart and fixed it, but from then on, I always looked over the side of the boat and estimated how fast we were going. I also tried to orient my senses in other ways, noting the position of the stars, moon and planets at night, the direction of the wave train, and the bearing at which the sun rose and set each day.

It was a lesson that would stay with me for life, and provide much contemplation about how modernity and instrumentation can blunt our senses. It was something that David Lewis and I had talked about, and he'd written about in *We, the Navigators*. Bernard Moitessier also wrote about it. In his sailing school, he made his students sail without a compass. More recently, Jack Lagan has written an excellent book on the topic, *The Barefoot Navigator*. Staying oriented is vital, and good fun.

After the brief lull near the reefs, the wind returned with renewed vigour, and we began to cover up to 145 miles a day. With the mainsail squared right out and the jib sheeted flat, *Ice Bird* thundered along, day and night, steered by the Aries self-steering gear. The boat, being slack-bilged, rolled a lot, with a distinct pattern. Each roll to leeward would be a little deeper than the last, and the boat would take more time to roll back to windward, so the roll period became longer and longer.

*We rolled from side to side like this for 19 days. It got quite tiring.*

Eventually, after three or four rolls, *Ice Bird* would not get back upright before the next wave exploded under the chine. This sent the boat into a really deep roll, submerging the leeward sidedeck, and it would go into a bit of a broach. The Aries would be flapping on the stern like a seagull fighting over hot chips, but it always brought us back. Initially, we ran up on deck when *Ice Bird* broached, grabbing the tiller, but after a while we just let the boat get on with it. We came to recognise when it was going to happen, and just hung on.

*We were really hammering along across the Coral Sea, averaging over 140 miles a day.*

Before long, we were approaching Mellish Reef, between New Caledonia and Australia, in the middle of the Coral Sea, and once more we were scheduled to pass it at 0300. I altered course to keep the reef 30 miles off the beam. I knew from my extensive reading that this was the recommended minimum. Ocean currents are notorious for setting in unexpected directions near reefs. Satellite navigation has changed this. 10 miles would be acceptable now, even if the chart datum was inaccurate, as it sometimes is, though I suspect I would still give myself a little more room. 10 miles on a dark, windy night still seems too close to me.

The next day, there were several white-hulled fishing boats around, about 70' long, steel by the look of them. They were the only ships sighted during this passage, apart from the one that nearly ran us down on the New South Wales Coast, and another on the morning we cleared Frederick Reef. The sea was otherwise endless and empty, wave after indigo wave rolling by, white-caps as far as the eye could see, apart from occasional dolphins or flying fish. The flying fish that landed on deck, while numerous, were only about 2” long, hardly big enough to make breakfast with, as my heroes wrote about. By the time we found them, they had dried out like potato crisps, but we were not tempted.

*The swell built up to impressive heights.*

*The Aries windvane steered Ice Bird day and night without fail, even if it did flap like a fractious seagull at times.*

I spent a lot of time staring at the sea, thinking of my heroes and their stories. Many of them claimed to have developed a harmony with the ocean, to have been endlessly fascinated by life at sea. I was in the groove, for sure, and would happily have continued on around the world with Ros and Gordon, if that had been our intention, but it was the lure of landfall I looked forward to. Perhaps, if we had continued on around the world, I would have discovered this harmony in due course. It reads well, very poetic, and I'd like to think it was true. I still long for it, that serene place where you are at the centre of your universe. I have only ever glimpsed it.

17 days out, we passed east of Rossel Island, the easternmost of the Louisiade Archipelago off the southeast tip of Papua New Guinea, once more without seeing it. The next day we passed Pocklington Reef, still without visual contact, and entered the Solomon Sea. The Coral Sea was behind us.

*In the approaches to Pocklington Reef, the seas became very steep, causing me some anxiety.*

In our approaches to Pocklington Reef, the seas became very steep, and changed colour from indigo to bright green, giving us much anxiety, but soon afterwards the swell calmed down. By noon we were gliding across a flat sea for the first time in three weeks, and the sensation was electric.

The wind was easing, too. *Ice Bird* slowed, but we were in no hurry. We only had a few hundred miles to go now, and we all relaxed. Food was the only issue. We had plenty, enough for another three months or so, but quality was another thing. Not that Gordon or I dared say anything. Ros had wanted us to accompany her to the supermarket to stock up, but we'd declined.

'Well, don't you dare complain, then,' she said. She had done a good job, but all processed food begins to pall after a while. We began trying bizarre combinations, like tuna and jam, or peanut butter and raw onion. It was delicious. At least we had not been forced to eat the South African food that remained on board from Barry's Indian Ocean crossing. The most unappealing item, we decided unanimously, was a brand of slimy, colourless, flavourless sausages called *Weinerworsters.* Even the birds refused to eat them, squawking in outrage when we threw some over the side.

'They look like dead men's fingers,' said Gordon, amid much hilarity.

On the 18th night out, having sailed almost 2500 miles, we hoped to see the light at Cape St George, on the southern end of New Ireland, but failed to do so. We kept sailing, despite having run our distance, since we were only moving at 2 knots, there were no off-lying dangers, and the night was calm and clear. The wind had died to a whisper. Dawn revealed an empty horizon, and Ros and Gordon looked at me.

'We'll have to wait until I can get a position line,' I said, trying to sound more assured than I felt. At least the sky was clear and the sea flat. There was no point taking a sight too early, as refraction on the horizon distorts the results. I preferred to wait until 0900 if possible. Shortly before this time, a crack opened in the apparently clear sky, way up above the mast, to reveal a slice of jungle.

*Land ho.*

Within minutes, the haze had cleared, and we found ourselves craning our necks to look up at a huge mountain towering over us. We were only three miles off the beach of New Britain, near a place called Merai, I discovered after taking a morning sight, having sailed right past Cape St George during the night.

The wind had died completely, so we had plenty of time to examine our discovery. Only when we started drifting away from it did we become a little impatient. Never mind, we said to each other, the seabreeze will kick in soon. But it didn't. It was now mid-October, and the trade wind season had ended.

We were only 50 miles from Rabaul, but our tiny petrol engine, if it could be induced to run, which most of the time it couldn't, did not have enough fuel to cover that distance. We drifted for 15 days, first southwards towards the Trobriand Islands for 200 miles, and then east, until we were only 60 miles off Bougainville Island, which is administered by Papua New Guinea but is geographically a part of the Solomon Islands. Then we drifted north again, to exactly the same spot we'd started from.

*Becalmed in the Solomon Sea for 15 days, it felt like we were drifting through eternity. Here we have the jib up, trying to eke out a few miles, but the fickle breeze never lasted long.*

The sails lay in untidy heaps upon the deck. At first, none of us minded much. We read, played cards, stared at the horizon. The sea was as placid as a lake; there was not even the vestige of a swell, and the sun blazed down out of a cloudless sky. It was as if we were suspended in a vast, blue orb. At night, the stars were reflected in the mirror of the sea and it felt like we were drifting through eternity.

Once, we thought we could see a man sitting in an open boat. We all had a look through the binoculars. He appeared to be ignoring us. We decided that he could only be a shipwrecked sailor, this far offshore. *Was he dead?* It was a creepy thought, but Gordon coaxed the motor into life and we puttered over. It was a huge tree; the 'man' was the roots, sticking up in the air, and we laughed with relief. Tricky thing, the imagination, when urban constraints are absent. It wasn't the sort of thing you'd want to hit on a windy night, though, even with a steel yacht.

After several days, Ros and Gordon became edgy. They spent hours hoisting and dropping sails whenever the illusion of a breeze teased across the decks. It was never enough to keep the sails full for more than a few minutes. Interestingly, I was perfectly content. I'd found my oneness with the ocean at last. The others even accused me of not *wanting* to get there, because I wouldn't pull the sails up at every vagrant puff that came by.

And it might have been true. I certainly had no desire to return to Sydney, and was in better humour than I could ever remember. What a wonderful life. I'd long ago stopped wearing clothes. What Ros and Gordon thought of this they were diplomatic enough not to say, and I seemed content to drift forever in naked bliss.

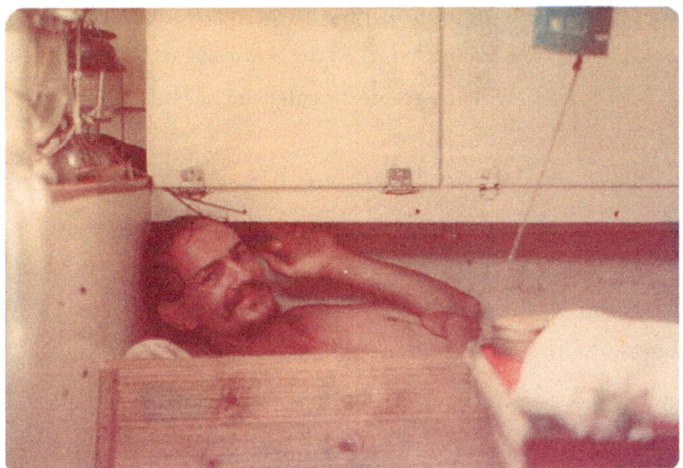

*I was content to drift forever in naked bliss. We all loved this 20" wide, port-side bunk with its solid timber leeboard. It was more restful to sleep in than the starboard bunk, which was 22" wide and had a canvas leecloth, which allowed the body too much movement as the boat rolled down the trades.*

I was so content that only the threat of dead men's fingers for dinner got me off my backside, though I must admit, I sometimes dreamed of an icy-cold drink with condensation dripping down the outside of the glass. But when, on the 15th day of calm, beneath the mountain we'd first made landfall on, *Ice Bird* began to drift south again, even I despaired. I spent the day on deck, tweaking the sails.

That night, a thunderstorm struck the boat while I was having a crap over the stern, holding onto the backstay. The world turned electric blue, the backstay sizzled like a hot-plate under my hand, and, simultaneously, a thunderclap heard in Hades shook the deck. I bolted below, not bothering to wipe my bum. To my astonishment, there was nothing wrong with my hand. This was followed, however, by a nice little breeze, and we all scrambled on deck to get *Ice Bird* moving. The wind was fitful, and from the northwest, which meant we had to beat into it, but we didn't care. Stagnation is death. Movement is the nectar of the gods.

The breeze had steadied by the morning, and on we sailed, holding our breaths. Up St Georges Channel we went, seeing distant waterfalls tumbling down lush, green mountains, gorging ourselves on the visual feast. By mid-morning, the Duke of York Islands were abeam, low on the horizon, mysterious and alluring. The temptation to slip in there was strong, but we resolutely turned the bow towards Rabaul Harbour. Besides the legalities of clearing in with the authorities, the lure of cold drinks and fresh food was irresistible.

After clearing customs, we stumbled ecstatically ashore. Flowering Hibiscus, Frangipani and Flame trees scattered a sweet-scented carpet of, yellow, white, and red petals underfoot, gardens were filled with blue and purple flowers, birds swooped across green lawns, while behind the town, the steep slopes of *Big Mother* loomed, its volcanic crater smoking ominously. It was some years before the volcano destroyed Rabaul.

We bought cold drinks, a slab of butter, and a loaf of fresh bread, sitting in the gutter to scoff it all. The locals laughed at us, their mouths, stained red by betel-nut juice, a startling contrast to their black faces. Our senses were rioting. This was the look I'd seen in the eyes of voyagers arriving in Durban as a boy, and it was addictive. I wanted to keep crossing oceans until I died, just to relive the rush of landfall.

The voyage on *Ice Bird* was a turning point in my life. It made everything that followed possible, even if there were still some hard times ahead. I will always be grateful to Ros and Gordon for being such great shipmates.

I was reluctant to return to Sydney, but there didn't seem much choice. On my way back, I passed through Mooloolaba in SE Queensland, which in those days was a sleepy fishing port, with lots of cheap moorings for ocean vagabonds who hung out there during cyclone season, and a beautiful beach across the road.

Daniel and Nicole Moulin, my French friends from Cammeray Marina, were there on *Henri*. They had been cruising the Queensland coast since 1974, and regaled me with stories of the Great Barrier Reef islands. I swore that one day, when my ocean voyaging days were over perhaps, I would base myself in Mooloolaba and cruise the islands of North Queensland. Not many of my ambitions have come to fruition but at least that one did.

From Mooloolaba, I travelled south to Brisbane, where Harry Gilbert had arrived from Hawaii on *Kelasa*. The boat was moored on the pile berths adjacent to the Botanical Gardens. After he had received a letter from Adrienne informing him that she was not returning, he'd sailed from Honolulu to Brisbane, with one brief stop in Vavau, Tonga, a marathon, singlehanded passage. He intended *talking sense to Dolly*, but Adrienne moved before he arrived. He was keen to hear if I knew where she was, and I lied, hoping I didn't look too guilty. Life, which had seemed so simple to my naive boyhood self in Durban, was proving to be far messier.

It was great to be back in *Kelasa's* cabin, with its huge deck beams, hanging knees and varnished mast dominating the saloon. It always reminded me of Joshua Slocum and *Spray*. The following summer, Harry sailed down to visit me, anchoring in McCarrs Creek, Broken Bay, 18 miles north of Sydney Harbour. The creek is behind Scotland Island, and in those days, it was a magnificent anchorage, sheltered from all wind directions. Visiting yachts often spent the summer there. Harry based himself in McCarrs Creek for several summers, sailing back out to Tonga in winter months.

Today, the creek is crowded with moorings, and anchoring is limited to 28 days a year in any one place, even if you could find a spot. Under the legislation, the whole of Broken Bay, with its myriad of bays and anchorages, is deemed to be one place, as is the whole of Sydney Harbour. Even if vessels move from one officially designated place to another, they are only allowed to anchor for a total of 90 days in a given year in the whole of New South Wales.

It was so much simpler back then, even if living aboard was technically illegal, as it had been for many years. There had always been a live-aboard community, as well as visiting yachts lying at anchor for the summer, and the authorities largely ignored the situation. When I used to ask what was wrong with living aboard, the reply was invariably that they did not want Sydney Harbour to turn into another Hong Kong. I thought a few junks around the place, filled with sea-gypsies, would make the place less boring.

During one of Harry's visits, I met an English couple, Tim and Pauline Carr, who were anchored near *Kelasa* in McCarrs Creek, repairing their 28' Falmouth Quay punt, *Curlew.* It was an engineless, 10-ton gaff cutter, built in 1895, and they were one of the happiest couples you could meet. They even finished each other's sentences. They'd found the boat in Malta in 1968, when Tim was 26 and Pauline just 21. It was very tired and destined for the scrapheap, but they were short of funds and looking for a cheap place to live. However, *Curlew* seduced them, and they had been voyaging ever since, working in various ports along the way to raise funds, repairing and rebuilding the boat as they went.

They had sailed to Australia via the Caribbean, Panama Canal, Hawaii, where they refastened the hull, and the South Pacific Islands. Over the years, they'd thrown out the engine, toilet, batteries and electrics, plus anything else that could lighten and streamline the ship. They got rid of all the baggywrinkle, the sidelights, ratlines, etc, fitted slim, stainless-steel rigging, and new, superbly-cut sails. Tim had even recessed the three remaining skin fittings, fitting rubber flaps over them to reduce turbulence.

*Curlew in Robinson's Cove, Moorea, Society Islands, behind Kathi 11. Curlew still had the high coachroof that was on the boat when they bought it. Photo: Kathi 11.*

Somewhere along the line, they started racing *Curlew*, and became very good at it, usually trouncing the local opposition on their modern yachts. They always donated the silverware back to the clubs who sponsored the races, since they had no space to carry them aboard. Besides, Tim was a terrier for evading unnecessary cargo. God knows what they thought of 22-ton *Kelasa*.

But this time the situation was more serious. One night when *Curlew* was lying quietly at anchor in Sydney Harbour, kerosene anchor light in the rigging and Tim and Pauline snugly asleep, they were run down by a trawler, skippered by a drunk, which stove in the boat's topsides and lifted off half the deck. Amazingly, *Curlew* didn't sink, and they quickly got the boat up onto a slipway, where they replaced the smashed hull planking and broken ribs.

After the hull repairs were complete, they gingerly sailed up the coast to Broken Bay, where they replaced the damaged coachroof and decks with a strong, flush deck. They lived aboard throughout the rebuild, and for all of the 28 years they cruised *Curlew,* even in South Georgia during the winters, with three feet of snow on the decks.

*Curlew showing off its new, flush deck. The boat is now a living exhibit at the National Maritime Museum in Falmouth, UK. Photo: Falmouth National Maritime Museum.*

After completing deck repairs, they cruised south to Tasmania, where they worked to raise funds for two years and cruised southern waters. It was becoming obvious, however, that the hull needed serious attention. Tim was spending more and more time cutting out rot around fastenings in the planking and epoxying in dutchmen, or small pieces of timber, like Keith Kibler had done all those years before in Durban to *Korsar*. Being iron fastened, the planking was suffering from iron sickness, where the timber around the fastenings suffers delignification and begins to rot.

In 1975, while they were repairing *Curlew's* planking in Honolulu, John Guzzwell (who was there on *Treasure*) suggested that they should sheath the hull in timber veneers set in epoxy, but they'd been worried about adding weight. Now, it was becoming obvious they would have to do something; sell the boat, re-plank it entirely, or sheath it.

An article in *WoodenBoat* magazine, about sheathing old wooden hulls with epoxy and cold-moulded timber veneers, laid to rest their fears about making the boat too heavy (it actually ends up lighter and more buoyant), so they decided to sail across the Tasman Sea to New Zealand and give *Curlew* the same treatment.

In late 1984, they set the boat up ashore in a shed so that it could dry out for a couple of months, while Tim countersunk the old fastenings, a time-consuming task, and faired the rough old pine planking. *Curlew* was not looking at its best there for a while, and they placed a large sign outside the shed door, which read: *Optimists only past this point*.

When the hull was fair, they sheathed it with three layers of 4mm kauri veneers, set in epoxy resin. They could only work when the weather was dry, so ended up putting in 20-hour days when conditions were suitable. The project, which took five months of solid labour, was a resounding success, giving *Curlew* a new lease on life. Tim's only regret later was that he did not protect the hull with fibreglass cloth after adding the Kauri laminations.

The hull was now as tight as a drum, allowing them to tension the rigging more than previously possible, which rewarded them with even more silverware and blisteringly fast passages.

After sheathing the hull, they made an incredibly fast passage across the boisterous trade winds of the Indian Ocean to Mauritius and South Africa, averaging 6.82 knots for 2500 miles at one stage. They made the fastest port-to-port passages of all the yachts crossing the Indian Ocean that year. They then sailed back to Caribbean, where they crossed their outward track, completing *Curlew's* circumnavigation, triumphed in all the local regattas, including Antigua Week, worked for a year, and sailed on to Maine. After wintering-over there, they crossed the North Atlantic to Falmouth, where *Curlew* was built almost 100 years earlier. After another year working ashore, and trouncing the local racing fleets, they sailed south to the Falkland Islands and South Georgia, visiting the Antarctic Peninsular while there.

They ran the whaling museum in Grytviken, South Georgia, for several years, living aboard through Antarctic winters, with snow piled up three feet deep on the decks, and wrote a beautifully-illustrated book about their time there, *Antarctic Oasis: Under the Spell of South Georgia*.

In 1991, recognising the Carrs' achievements, the Cruising Club of America awarded them the prestigious Bluewater Medal. They were also awarded the Royal Cruising Club's Seamanship and Tilman Medals, as well as the Seven Seas Cruising Association Award. They were revered in the voyaging community for having explored the remotest cruising grounds imaginable, over almost three decades, in such a small, simple, engineless boat. After donating *Curlew* to the National Maritime Museum in Falmouth in 2003, Tim and Pauline retired to a quiet life in New Zealand. Pauline died from pancreatic cancer there in August 2019, while Tim continues to lecture and conduct ecological tours.

*Pauline Carr rowing ashore from Curlew in sub-antarctic waters. Photo: Tim Carr.*

Another sailor I met at that time was Peter Morgan, who had a waterfront house opposite Mitchell's Marina and an old William Garden ketch that had been built in Asia. He complained about the poor workmanship that had gone into the vessel, and was busy tearing it apart. I got to know Peter well in our mature years, and realised that he was never satisfied with a boat until he'd taken it apart and put it back together the way he preferred. It was something he'd done with dozens of old classic yachts. He is one of those guys who likes to keep his hands busy. He was friends with the Carrs, who were anchored near his house, and still speaks of them with great respect. I'll return to his story later.

I developed the habit of visiting Harry every Wednesday night, taking a basket of food and wine. We'd have a feast and talk about voyaging until the early hours. I never tired of listening to his stories, even if they tended to go on a bit. Then I'd get on my small motorcycle and weave my uncertain way back to my shack along the back roads of the Kuringai Chase. I only crashed the bike once.

One Wednesday, *Kelasa* was on the slipway at Cottage Point, getting a fresh coat of antifouling paint, but we'd agreed to meet as usual that evening. Early that morning, Harry rang me at the boatyard I worked in, asking if I'd like to come down and help him with the painting.

'Gee, I'd like to, Harry,' I said, 'but I can't just take a day off work without notice.'

'Oh, that's OK,' he replied, sounding relaxed. 'I just thought I'd ask. I'll see you tonight.'

I never thought any more about it. When I arrived, he sat me down in my usual place, wedged in beside the mast, the same spot he had always instructed me to sit, ever since I began visiting him in Dead End Creek, Durban. I have seen pictures of David Matzenik sitting there too, aged 17, in 1967. It was where, it seems, Harry always placed his visitors. It was also where he sat when he was alone on the boat, so I suppose I was the guest of honour.

Harry stood by the stove, preparing the meal. 'Tell me,' he said, stuttering, something I'd never heard him do before, 'Do you ever have bad thoughts?'

*Oh no*, I thought, *not another middle-aged seducer*. There had been several others recently, some quite unlikely, married friends and the like. My little-boy-lost persona was irresistible, it seemed. I'd begun to wonder about Harry, who had been showing signs of emotional dependency.

*David Matzenik, aged 17, in Kelasa's saloon, sitting in the same place Harry always put me, wedged between the mast and his bunk. It was where he usually sat, so I think it was a concession to the guest of honour. Photo: courtesy David Matzenik.*

'What sort of bad thoughts?' I asked reluctantly.

'You k...k...know,' he stuttered, 'r...r...*really* bad thoughts.'

I shifted uncomfortably in my seat. 'If you can't be more specific, I can't say,' I said, my scalp beginning to prickle as I remembered Adrienne's stories. Harry's position by the stove blocked access to the companionway. There was no other quick exit. He had a strange look on his face.

'Well, just to give you an example,' he said, 'you ring a friend up and ask him to help you paint your boat, and he says he can't, and you know he has a perfectly reasonable excuse, but you spend the rest of the day thinking about cutting his head off with your axe.' His axe, intended for cutting the mast away in a hurricane, was just behind him, strapped to the bulkhead.

I'd never known whether to believe Adrienne's stories. She loved a good yarn, especially over a glass of wine or two. It was an old sailor's tradition, shades of *Moby Dick*, to tell tall tales, but now I suddenly remembered something else she'd told me. *You have to be firm with Harry,* she'd said, *show superior mental strength, stare him down and shout.*

'No, I don't have thoughts like that, Harry,' I shouted. 'It's totally unreasonable.'

'Yes, of course it is,' he said meekly. 'Now, would you like a glass of wine?' And the evening proceeded as normal.

Harry sailed back to Vavau, northern Tonga, that autumn. *Saving the cost of a winter overcoat*, he said with a knowing smile, quoting our mutual hero, Joshua Slocum. From there, he returned to Sydney in the summer months, repeating the voyage the following year. That became the pattern of his existence for several years. Some summers he went to New Zealand or Tasmania instead, but always stayed in touch, and eventually returned to McCarrs Creek. If he had once found *Kelasa* difficult to manage alone, as he told Adrienne when they first met, he seemed to have worked things out.

*Kelasa steaming down the Hawkesbury River, Broken Bay, in 1978.*

His passages were outrageously slow, especially after he was caught by a squall with full sail up one night south of Tonga. The weather had been quiet, with moderate seas, a sky full of stars, and a lovely following breeze. *Kelasa* was broad-reaching with the mainsail squared right out, the boom preventer snugged up, the boomed staysail sheeted to the tiller, keeping the boat on course, and the jib, at the end of the 12' bowsprit, sheeted tight.

He came up on deck for a pee at about 0200, and saw a tiny cloud to windward, but apart from noting that it was a peculiar brown colour in the moonlight, looking like a stone, he said, he thought little of it. He went back to sleep and woke to a screaming squall, with *Kelasa* hove down to a crazy angle. Rushing on deck in his underpants, he took the tiller and drove the ship downwind for several hours, soaked and freezing, despite the tropical latitude. Then, in a lull, he cast off the halyards, climbed the ratlines, and jumped up and down on the gaff to bring the mainsail down.

After that, he rolled up the jib and double-reefed the mainsail before retiring in the evenings. *Kelasa*, effectively hove-to, jogged through the night while Harry got his famous eleven hours in the bunk. It was not unknown for *Kelasa* to take 50 days to sail 1000 miles. In gales, he took all sail down and drifted.

Because *Kelasa* had no electrics, the ship was unlit most nights, though near shipping lanes he hung a kerosene hurricane lamp on the boom gallows, for what it was worth. When *Vivienne*, the ancient, unreliable, petrol engine, was replaced in Hawaii with a diesel motor, he chose not to install a battery or electrics. The boat had Dacron sails now, at least, so he was not interminably stitching half-rotten flax.

He navigated by sextant to the end. Naturally, there was no safety gear on the boat, no radio, flares or liferaft, not even a harness to tether himself. Mind you, *Kelasa* sailed so slowly most of the time that he could have swum back to the boat. He was no stranger to solitude.

I think I was his only friend in those days. He wrote me very long letters while at sea, 30-40-page missives of interminable detail. I loved receiving them and remember their beautiful stamps. The Tongan ones were shaped like bananas. Once, 42 days out of Sydney, while beating painfully to windward off the west coast of New Zealand's North Island en route to Nelson, he wrote, somewhat outrageously, that he was starting to like ocean sailing. You just have to stick at it, he said. He'd sailed more than 70,000nm at that point.

In all, he sailed over 100,000nm, and never got into any sort of trouble at sea. He may have been half-crazy, but he was also a superb seaman. The other thing is that nothing ever broke on *Kelasa* in all those miles, despite enduring numerous gales, including one survival storm in the Tasman Sea in 1967. In that incident, tropical cyclone *Gizelle* came down from the Coral Sea and collided with a deep low. It led to the sinking of the 8948-ton passenger ferry, *Wahine*, with the loss of 53 lives, in Wellington Harbour, New Zealand. The wind-strength was recorded at over 100 knots, and *Kelasa* was not far to the west of the storm's path.

*Kelasa on the hard at Mooloolaba, Queensland, Australia.*

Harry never claimed a love of the sea, saying he was an escapologist, and that he voyaged in the hope of finding something, somewhere, though he wasn't sure what.

# Chapter Thirteen

# The Carousel of Life

After I returned from Rabaul in 1976, I sold the uninspiring ferro-cement hull for less than the materials I put into it, and bought a small fibreglass hull and deck from Barry Lewis. It was a Hunter 19, designed by Oliver Lee, a talented English designer, with 1500lbs displacement, half of which was in the keel. These boats had a reputation for offshore work; David Blagden sailed one, *Willing Griffin*, in the 1972 OSTAR, and they had the potential to go almost anywhere, although being a Hobbit might have helped. It was good that lying down was one of my favourite occupations.

The Hunter 19s that Bill Norman built in Sydney were all solid, with excellent attention to detail, but Barry specified significantly stouter scantlings, as he planned to take the boat around Cape Horn. I wasn't that daft (nor was Barry after sailing *Ice Bird* from Cape Town to Sydney), but I figured it would make a fine tropical trade wind voyager. The Hunter 19 is similar in proportion to *Sopranino*, the progenitor of *Trekka,* though arguably somewhere between the two designs, having a little more beam and displacement than *Sopranino*.

It is probably stronger, too, though *Sopranino* was brilliantly engineered, and one of the true pioneers of light-displacement boat design and construction. Patrick Ellam and Colin Mudie sailed 19' *Sopranino* across the North Atlantic in 1952, and their book recounting that voyage is a delightful read, especially those editions with the original whimsical sketches. They made it sound like so much fun. They even had a stuffed elephant aboard, Hannibal, who had his own hammock, and was accused of drinking all the booze.

*Patrick Ellam, left, and Colin Mudie (with tie!) aboard Sopranino during their transatlantic voyage. Photo: Sopranino archives.*

So, it was back to the old dream, the smallest, simplest boat I could squeeze into, etc, etc. I named the boat *Adrienne Matzenik*, hoping that some of Adrienne's feisty spirit would rub off. I made a mess of the fit out, in my usual impatient manner, but there is no doubt that it was all strong enough. I had a seaworthy yacht at last, and the islands of Queensland, and beyond, beckoned.

I wanted to give *Adrienne Matzenik* a junk rig, but Bill Norman was opposed to the idea, saying it would be too heavy. I think he was unaware of the trend towards alloy masts and was thinking of solid timber. He changed his mind later, when someone else ignored his advice, as I should have done, and fitted junk rig to their Hunter 19.

But I can't blame Bill for the unfortunate need I had in those days to please people, especially older men. Father complex, perhaps. Bill discovered that the extra weight of the junk rig improved the boat's motion at sea, given it was designed with a 50% ballast ratio, and the standard boat's motion was rather quick. I might point out that weight aloft in light weather is never a problem, and when the junk sail is deeply reefed, the centre of gravity comes down with the yard, and is then probably lower than it is with a stayed bermudan rig.

David Lewis also advocated against junk rig, which surprised me, as he was the first person, in 1966, to tell me about *Jester* and enthuse about the concept. I reminded him of this. 'Yes,' he said, 'but since then, a number of them have broken their masts. I am not sure if unstayed rigs are a good idea.'

'But David," I said, amused, 'you've broken your mast at sea about four times now, haven't you? Once on *Cardinal Vertue*, once on *Rehu Moana,* and twice on *Ice Bird*.'

'Oh, gee,' he said, looking a little abashed.

But I bowed to their wisdom, or lack of it, and rigged *Adrienne Matzenik* as a bermudan sloop, with twin forestays and jib poles, for running under twin headsails before the trade winds, modelled on Julio Villar's *Mistral*. I hoped to escape Sydney in the winter of 1977, sailing north to the tropics at last, but wasn't quite ready. Instead, I accepted an invitation to navigate a 10-ton, 36' ketch, *Silver Heels,* to the Whitsunday Islands for George Bennett, whom I'd met when the boatyard I worked in built his yacht.

*Silver Heels* was a John Sampson C-Shell design, with a ferro-cement hull and teak superstructure. It had all the basic comforts, including a powerful diesel engine, and was a luxury yacht compared to *Ice Bird,* an ocean liner alongside *Adrienne Matzenik.* It was still a simple boat, without refrigeration or electronics. We had an almost effortless, incident-free trip, a demonstration of the power of practicality. I recognised this, but despaired of the cost and time required to acquire such a boat.

*Silverheels 11 in McCarrs Creek, Broken Bay.*

In Mooloolaba, I discovered that Daniel and Nicole had sold *Henri* and moved back to France. I sadly fingered the shell necklace they'd given me, which I always wore like some talisman of a better future. I had been so looking forward to joining them in the islands, and I was deeply unsettled by their decision to move ashore. It made me feel vulnerable and afraid in ways I did not understand.

Just north of Great Keppel Island, we had one of those experiences that remain with you forever. There was a low-pressure system down south, blocking the ridge of high pressure that fuels the SE trade winds on this coast, so the wind was almost non-existent and the sea flat. I was steering *Silver Heels*, motoring quietly along, while George sat in the cockpit opposite me, enjoying a pleasant yarn. Suddenly, he leaned forward, shoved the tiller into my stomach, and started jabbering incoherently, it seemed to me, about killer whales. I thought he'd had a stroke or something. I pushed him back into his seat, yelling for our third crewmember to come on deck.

'Killer whales, killer whales,' George kept repeating, then pointed to port. Up alongside the boat, about a boat-length off, came a huge shark, a tiger shark I think, maybe 6-7m long. We were motoring at about six knots, but it was doing at least 10, hardly making an effort. Its tail was languidly swishing, and its head twitched from side to side. Then, when amidships, it turned towards us and charged, immediately turning into a blur of movement. At the last moment, it dove, and its grey back turned to white underbelly as it disappeared under the keel. There was no impact, but a few minutes later it did it again, then we saw no more of it. No swimming today.

*Silver Heels* spent a day at West Bay on Middle Percy Island, where the sole permanent resident, Andrew Martin, had built an A-frame shed on the beach for yachties to hang out, buy honey, eggs and goatskins, and shower beneath the rainwater tank. The bay was totally swell-free, making the anchorage a vision of paradise. I did not realise, having never been there before, how rare those conditions were.

There was a 55', three-masted, Herreshoff-designed, Marco Polo here, *Otter*, and the black-bearded lone skipper, John Ashford, regularly swam around the bay visiting yachts. I told him about the huge shark that menaced our boat just north of the Keppel Islands, but he just laughed. I later heard that West Bay is renowned for sharks.

The A-frame shed was filled with placards from visiting yachts, and there was often an interesting gathering of sailors within its shady environs. I was excited to see Tom Corkill's sign, recording past visits, with a recent addendum indicating that he'd been there on *Ninetails* a few weeks earlier. I hoped we might catch up, but learned later that he had sailed on to Thailand. In Brisbane in 1975, he'd told me he was through with long-distance voyaging, that it had become too crowded and expensive, but it seems the lure of adventure and new horizons had snared him once again.

There was an older shed nearby, erected by the previous occupants, the White family, in which I found placards and visitor's book entries from earlier voyagers. On a piece of tin, nailed to a post, John Guzzwell had painted *Trekka's* name in red paint. It was a little faded after 20 years but still legible, and the Hiscock's sign for *Wanderer III* was nearby. Keith Kibler's placard was there too, the name, *Korsar,* neatly-stencilled name standing out in bold, cream paint, looking like it had been painted yesterday, while Croix's sign had faded badly, and one could just make out the words *Iorana, Auckland, 1969.*

*Keith Kibler's neatly-stencilled sign.*

So many memories. I dug my toes into the hot, white sand, trying to ignore the sand-flies biting my ankles, and squinted through the palm trees at the azure ocean and the yachts anchored in the bay. I was here now, but why did I feel like a fraud? I had not yet heard of impostor syndrome.

From there, we sailed to Scawfell Island, a pristine National Park zone, where the deep, U-shaped anchorage is fringed with coral, and the beach ashore with leaning palm trees. Apart from the wind in the trees, and the sound of bird-call, it is a place of deep, exquisite silence. I had wanted to visit this island since reading about it in Lawrence le Guay's book, *Sailing Free*, and it did not disappoint.

All too soon, we arrived at Airlie Beach, where I had to leave the yacht, as George's family were due to arrive in a few days. At Middle Percy Island, I had been invited to join a yacht sailed by a man and his teenage sons once they got to Airlie. The boys were great fun and I really wanted to spend a few weeks up there with them, plus the islands looked so enticing. However, I was short of money, as usual, and the trip on *Silver Heels* had already set my project back another year. Regretfully, I returned to work in Sydney.

*Adrienne Matzenik's* completion was further delayed by my involvement in David Lewis's latest expedition the following summer. He planned a return to the Antarctic, this time with several scientists as crew, on the ex-racing yacht, *Solo*. I'd had so much fun working with him on *Ice Bird* in 1972 that I could not resist, despite the delay.

*Solo* was a steel yawl, 57' on deck, with a beam of 13' and a draft of 8'. It was designed by Alan Payne and built by the owner, Vic Myers, in Sydney in 1955. The boat was intended for cruising but was so fast that Vic took it ocean racing, winning 80 races over the next 8 years, including four Sydney to Hobart races, never finishing in less than 4$^{th}$ place. *Solo* was low-wooded (if you can say that about a steel boat), and wet, so much so that the crew were frequently washed off the foredeck when changing headsails. When they complained, Vic told them to toughen up, but when he later had to do it himself, he fitted a solid pulpit in the bows, nicknamed the cow-catcher.

When *Solo's* racing career ended in 1963, Vic went cruising, first sailing around Australia alone, then embarking on three successful circumnavigations with crew, often female. He became a bit notorious for this. In Tahiti, on his last voyage, Vic hung a 'crew wanted' sign on the stern rail, and some wag added the letter 'S' to the front of the word 'crew' one night. Vic, we were told, didn't notice for a day or two, to the sniggering amusement of other sailors on the Papeete waterfront, but when he did, he took the sign off and threw it into the lazarette, where we found it under some coils of rope.

*Solo* was a large yacht but still relatively basic. It had a radar and powered winches, but not much else, and there was a lot of work to do to accommodate a crew of eight for a three-month voyage to the ice. Much of the refit took place at Dangar Island, where David and Yvonne had set up house. This island on the Hawkesbury River remains one of my favourite places, beautiful and serene, though life in David's household was anything but at times, overrun by expedition crew and support teams. David's partner, Yvonne, who was doing a good job managing logistics, was a territorial soul and not coping well with the intrusion. The atmosphere in the house was frequently fractious.

It was a very different experience to preparing *Ice Bird* in 1972, when essentially there was just David and me. Trying to please several headstrong expedition members exacted its toll on David. They were becoming mutinous, and were very competitive when it came to allocation of space aboard for their pet scientific projects. David was often irascible during the day, but reverted to his usual, colourful self at night, after a couple of rums.

One morning, when he was agonising about his crew, I said, 'David, a sailing ship is not a democracy.'

That evening, when we gathered for drinks, he shared my wisdom with the crew, but only succeeded in making both of us unpopular. The boat hadn't left the dock yet and it had already become a pressure cooker. I was glad I wasn't going, as David had originally suggested. Besides shying away from confrontation, I hate the cold.

Nothing could deter David, though. Sometimes, in everyday affairs, he came across as comically flustered, but behind that modesty was an unshakeable determination. He throve on adversity, something that Blondie Hasler recognised in the lead-up to the 1960 Singlehanded Transatlantic Race. After Blondie met David, he said that he *knew* there was going to be a race. *Solo* would get to the Antarctic, and they would achieve their objectives. Anyone who thought they could bully David would ultimately fail.

One incident showed David's character in its truest light. We had backed *Solo* into a berth alongside a jetty at high tide, as there was not enough depth there at low water, to do some work that required shore power. Inevitably, work was delayed and the keel began bouncing on the bottom. Time to go. But the diesel wouldn't start. Some of the crew were living aboard, and they'd flattened the starter battery, which was supposed to be isolated from the house batteries, but hadn't been.

We rowed out an anchor and attempted to pull ourselves off, but *Solo's* powerful hydraulic anchor winch was useless without the motor. We only succeeded in moving the boat a few feet, just far enough so that it would no longer lie against the jetty when it dried out. The mizzen mast would now come down across the jetty, probably causing damage worth thousands of dollars. Given the usual fiscal and temporal limitations of amateur expeditions, this would most likely have scuppered the voyage.

So why was David, who'd been grumpy all morning, now grinning like a five-year-old at a birthday party? He already had a plan, to undo the mizzen mast's rigging and lower the stick onto the jetty, utilising the mainmast halyards, and looked all set to relish the opportunity for some theatre sport. He would have handed out scrubbing brushes for a quick bottom clean at the same time, no doubt.

Just then, Ces Turner, who owned the property, came running along the wharf with a battery charger. This was plugged in and gave the batteries just enough life to fire up the diesel. The combined grunt of the diesel and the winch, hydraulics now restored, dragged us off in the nick of time. David remained up-beat all afternoon.

'So,' said Fritz, the charismatic joker of *Solo's* crew, a short, stocky man with an unruly tangle of black hair, 'now we know what we have to do to cheer the old bugger up. Just put a hole in the hull.' It was to prove a prophetic statement, as *Solo* did end up with a small hole, courtesy of the pack ice and a rusty spot in the shower cubicle. David gleefully reported afterwards that Fritz had not been so amused then.

David came back from the voyage with an endless fund of Fritz stories. He claimed that they didn't need a compass. All they had to do was turn the boat north and Fritz would smile! Fritz was also notorious for his pranks during the voyage, such as cutting the bottom out of the pee container that David used at sea.

Perhaps understandably, David suspected that it was Fritz who kept putting wet gloves inside his sleeping bag on the homeward passage. David would take his clothes off and wriggle into what he hoped was a warm, dry cocoon, then shriek with disgust as he came into contact with the cold, slimy gloves. Tearfully, he'd plead with Fritz to stop, but Fritz denied responsibility.

'Silly old bugger is losing his marbles,' he'd chortle. 'Puts them there himself, I reckon.' The culprit was never revealed. Possibly David did leave them there himself, to dry out, but he vehemently denied it.

A few days before departure, when *Solo* was on public display at Circular Quay near the Sydney Opera House, we had a visit from Val Howells, who'd been a fellow-contestant of David's in the inaugural Singlehanded Transatlantic Race in 1960, sailing a clinker-planked Folkboat, *Eira*, with an oversized bermudan sloop rig. Chichester had claimed to be worried about the *black-bearded Viking* winning the race, despite competing in his 40' sloop, *Gypsy Moth III.* Val wrote an unusual, evocative book about his voyage, *Sailing into Solitude*, which I'd read several times in the early days in Durban. Val was now sailing around the world alone on *Unibrass Brython,* an engineless, 38', flush-decked bermudan sloop he'd built himself, initially for the 1976 OSTAR.

*Val Howells on the deck of Unibrass Brython.*

There was no electricity aboard, only oil lamps. 'Just as I like it,' Val said. He'd suffered a head injury in the 1976 race and had to retire. Feeling depressed afterwards, he decided the cure would be a solo circumnavigation. Although 53, he was a tall, strapping man (6' 3"), with a luxuriant grey beard and a lion's laugh. He'd arrived in Sydney with just A$18 in his pocket, but soon found work. His daughter lived in Sydney, so he planned to stay for a while.

I was amused by David's reaction after Val left. Looking somewhat mortified, he said, 'In 1960, I was the fit and muscular one, and he was a bit of a slob. There was no competition when it came to impressing women. Now here I am, flabby and unfit, and he really does look like a bloody great Viking,' referencing Francis Chichester's famous quote about Val.

I remember being quite shocked by how much David had aged in the six years since the *Ice Bird* voyage. He'd been so fit and muscular in 1972, easily convincing female interests that he was 10 years younger than he actually was. He'd fallen and broken a hip while snow skiing soon after his return from the Antarctic, and the lack of mobility had taken its toll.

Despite successful hip surgery a year or so later, and a return to an active lifestyle, he never regained his earlier level of fitness. He had passed one of the staging posts of life, it seems. It is something I have witnessed in a number of friends, such as Harry Gilbert.

At the time, I assumed lifestyle had something to do with it, but understand better now, having turned 72 in 2024, that the body's self-regeneration powers, something you take for granted in youth, diminish with age. Not that David allowed physical infirmity to shape his ends. The wind might call the tune, as sailors say, but David danced his own merry jig.

*David Lewis taking a hurried sunsight in the Southern Ocean in 1978 aboard Solo. Photo: David Lewis archives, courtesy Barry Lewis.*

Another surprise visitor to Circular Quay was John Murray. Since I had last seen him in 1972, he'd completed a 7-year circumnavigation on *Unbound.* I knew he'd returned, because one morning in 1975 I'd opened the *Sydney Morning Herald* newspaper to find a picture of him rowing ashore in Watsons Bay, then the official anchorage for small vessels clearing Customs.

When I left Durban, he gave me his mother's phone number, saying I should call her, but cautioned that she was very reserved. Being shy, I had never called, but after seeing John's photo in the newspaper, I rang the number from a public telephone booth, not having a phone of my own. John answered the call, to my surprise. He was pleased to hear from me, but said he was very tired, suggesting I ring back in a few weeks. It never took much to deter me, and I never called again, so it was great to finally catch up with him.

It was a joyous reunion beneath the lights of the Sydney Harbour Bridge and the Opera House. Rum and laughter flowed into the small hours. David and John had much in common. They were both daring sailors who had circumnavigated the world on multihulls, and shared an irreverent sense of humour.

*Solo in Commonwealth Bay, Antarctica.*
*Photo: courtesy Barry Lewis.*

*Solo* sailed south to Commonwealth Bay in the Antarctic and back, with some dramas but no major catastrophes. The worst incident was the famous hole in the hull, quickly repaired with cement and a patch cut from a wetsuit. They endured a few serious gales, maybe not as fierce as the storms that overwhelmed *Ice Bird*, but perhaps it was also because *Solo* was so much larger, and arrived in the Antarctic later in the season. They hove-to under trysail, with the bows into the weather and were never in any serious danger.

*David Lewis at the helm of Solo in Antarctica. Photo: courtesy Barry Lewis.*

I asked David how he thought *Ice Bird* would have fared hove-to under trysail, rather than running off at speed, as he had done, but he felt that heaving to worked better in larger vessels. He said he had tried it in the North Atlantic on *Cardinal Vertue*, on his tempestuous autumn passage back to England in 1960, but the boat had sheered about, falling off at times so that it was beam on to the seas, and at risk of getting knocked down by breaking crests, then luffing at others, which threatened to shake the mast out of the boat.

To avoid sheering around like this, Larry and Lin Pardey set a parachute drogue off the bows of their small vessels when hove-to under trysail in severe weather, with a bridle to the windward stern quarter. Would that have worked for *Ice Bird*? Every storm, and vessel, is different, and the parachute drogue requires a fair bit of foredeck work, not so enticing in the Southern Ocean.

Perhaps a 5/8" diameter, 600' warp, towed astern in a bight, with both ends secured inboard, and a tiny storm jib sheeted flat fore and aft, would have worked best, like Robin Knox-Johnston's set up on *Suhaili* during his non-stop circumnavigation in 1968-9. But

*Suhaili* was a very different type of boat to *Ice Bird*, with heavy displacement, Norwegian stern, and a long keel.

I may have had some responsibility for David's decision to run off at speed in *Ice Bird*, as I loaned him my copy of Bernard Moitessier's *Cape Horn, the Logical Route*, and he read it while crossing the Tasman Sea after leaving Sydney. It was one of the few books that survived *Ice Bird's* passage to Palmer Station. I am a bit dubious about Moitessier's tactic for solo sailors who cannot steer continuously. *Joshua* put its keel in the air a few times.

A more recent development is the Jordan Series Drogue (JSD), a warp threaded with numerous small drogues and set off the stern on a bridle. It has been very effective on a number of small boats, including Noel Dilly's Twister-class yacht, *Bits*, and Roger Taylor's diminutive Arctic exploration yachts, *Mingming* and *Mingming II.*

Any sailor who has weathered a major storm at sea knows there is no definitive answer. Different ships, different long splices, as the old sailors put it, and the sea-state plays its part. On top of which, there is always an element of luck involved. I remember the wave I saw on *Poeme* in 1974, that climbed up the back of another until their combined height was almost double that of the surrounding seas, before collapsing in an avalanche of foam. If *Poeme* had been caught under that, we may not have survived.

*Adrienne Matzenik* was launched in April 1978, a few days before my 26$^{th}$ birthday. The intention was to do some quick sea-trials, then head for Queensland. Unfortunately, on the morning the boat was launched, I was knocked off my motorcycle by an illegal immigrant who did not speak English, driving a car with a forged license, who didn't know the rules of the road. The car struck me in the side, ran over me, and dragged me along beneath it. As I scraped along the tarmac under the car, the sound of my Marlin wet-weather jacket shredding sounded like a raw potato being grated, and my nostrils filled with the smell of burning rubber, petrol, oil and blood. I was sure I was going to die.

Surprisingly, my physical injuries were few, apart from a lot of grazed skin. I was soon back on my feet and sailing *Adrienne Matzenik* around the Hawkesbury River. There was no reason, really, why I could not have taken off for the islands of desire. However, the boy who crawled out from under that car was different to the one who got out of bed that morning. It is hard to rationalise, but it took a long time to recover mentally. As I was dragged along under that car, thinking I was going to die, I felt unbearably alone. The thought of sailing away on my own suddenly seemed appalling. *Adrienne Matzenik*, cute as it was, was too small, I rationalised, if you can use that word to describe my disjointed mental state, to take a crew. It was hard enough to squeeze myself into the tiny cabin.

*A couple of months after my accident. There really was no reason why I could not have set sail for the islands of desire. Photo: Tony Nelson.*

*Adrienne Matzenik on the Hawkesbury River. Photo: Tony Nelson.*

I decided once again that what I wanted was to build a steel boat. Harry designed the boat for me, a 28', single-chine, gaff cutter that was a streamlined, lighter version of *Kelasa*. It was a brilliant design, or at least it suited my needs and personality, just as *Kelasa* suited Harry. Given his anxiety, I doubt Harry would have gone to sea in anything less stout than *Kelasa*. There is no such thing as a universally ideal cruising boat. People rubbished the design when I showed it to them, and I joined in the fun by calling it *Eeyore*, after the sad-eyed donkey in A. A. Milne's *Winnie the Pooh* stories. The friend, who introduced me to *Winnie the Pooh*, Linnet Ewing, once said I reminded her of Eeyore. I would have liked a junk rig, but this bullet-proof boat, with its low-aspect gaff rig, which would allow me to climb the ratlines with ease, spoke to my core.

I moved ashore and rented a small property in Brooklyn, next door to a new friend, Terry McCartney, who was building a ferro-cement yacht in his backyard. Within weeks, I had sold *Adrienne Matzenik* to Tony Nelson, who sailed it around the Hawkesbury River for many years, and made a coastal voyage north to Ballina and return.

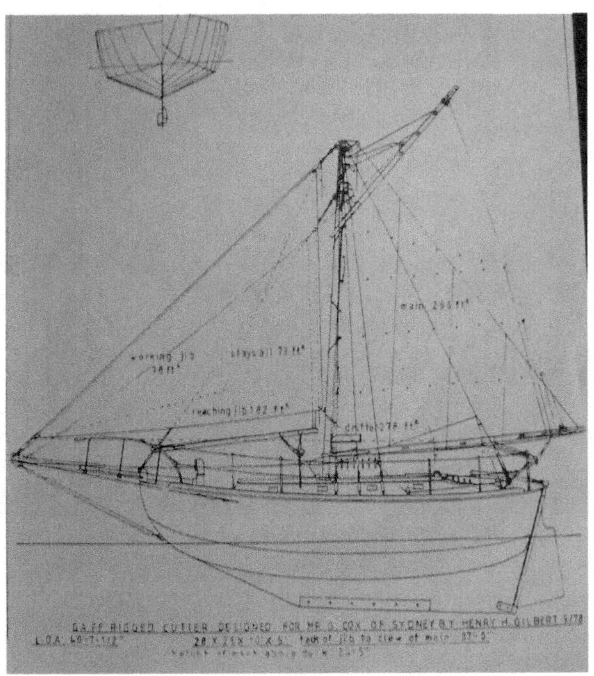


*The design I will always think of as Eeyore, the beautiful gaff-rigged cutter Harry Gilbert designed for me.*

In the end, I was talked into taking over the frame of another ferro-cement boat by Terry and a couple of others, who were also building boats nearby. They were worried that steel construction would be too noisy (it is), and cause trouble with the neighbours (it probably would have). I should have gone elsewhere, but I'd chosen that spot because Terry, and the merry band of hipsters who hung out at his place, were irresistible.

Summer came and went. In April 1979, I was 27 years old. Life really was getting away from me by this stage. Those magical days in Durban, when I thought I was about to leap feet-first into the voyaging world, seemed more distant than ever. Cooler weather, as usual, eroded my optimism, but I wasn't thinking of quitting, just lacking the raw enthusiasm that the right boat might have inspired. I was also dreaming of faraway places, palm trees and blue ocean. If I could just get back to Middle Percy Island, or to the Whitsunday Islands, I'd be ok, I thought. In the meantime, all I could do was cling to my painted pony on the carousel of life.

*Yet another damned ferro-cement boat, under construction on the property I rented in Brooklyn, June 1979.*

# Chapter Fourteen

# Bowen or Bust

Then, one bitterly cold day in June, Terry's phone rang (I didn't have one). It was Bill Norman, builder of the Hunter 19. Some bloke from Bowen, a tomato picker, Bill said, had bought a Hunter and wanted to sail it home, but had no experience. Would I spend a few days with him, sort him out if possible?

On the phone, the bloke sounded gruff. I imagined some rough 35-year-old with tattoos and missing teeth, who'd take one look at me, spit on the ground and leave. I was very much the hippy boy in those days, with long hair and embroidered shirts. Some of the local oyster farmers and fishermen called me a poofter, even though they knew nothing about my personal life, and they didn't much like that I'd painted my small cottage burnt orange. Altogether too colourful for their liking. Brooklyn was a working-class town in those days.

But when the car door opened and Rex Byrne stepped out, he was revealed to be a 17-year-old boy, in a clean white shirt and blue jeans, with unruly brown hair falling across his innocent-looking, freckled face, and a cheeky smile on his lips. He had his older brother, Tom, with him, who looked like he was appraising me. I felt like I was having an audition as we talked about boats and the sailing I'd done. I realised I must have passed when Tom suggested I sail with Rex all the way to Bowen.

At the age of 13, Rex had read Robin Lee Graham's book, *Dove*, and like me, decided he wanted to sail around the world. In some ways he was as naïve as you'd expect a 17-year-old to be, but he was also smart. He'd arrived in Sydney with A$5000, saved from his wages over the previous five years. Numerous yacht brokers tried to relieve him of his money, and themselves of old hulks they could hardly keep afloat, let alone sell. He held out, however, until he found *Mistral*, which had been fitted out immaculately by a man who dreamed of offshore sailing. It was the best boat his money could buy.

*Teaching Rex how to sail. You can see how fast Mistral is in this photo.*

10 days later, on 1 July, the little orange boat flew down the Hawkesbury River in bright sunshine before a 20-knot NW wind, bound for Bowen. We'd spent that time finalising the fit-out, teaching Rex a few basic sailing skills, and getting to know each other. We quickly formed a bond that would strengthen as the weeks passed.

*The day we left. Photo: Terry McCartney.*

Terry Macartney, and other friends, accompanied us down the river in an old launch, the *Flying Zucchini*, so named because it delivered vegetables to settlements upriver without road access. They had to open the throttle wide to keep up, but took some great photos.

*We were in high spirits as we sailed down the river. I am fooling with the sextant. Photo: Terry McCartney.*

Unfortunately, someone found a bag of raw potatoes aboard, and they began throwing them at us. 'Spare provisions', they laughed. Raw potatoes are hard though, and after one nearly smashed the face of the new speed/distance log display, I put my hands over the instrument to protect it. A potato soon bounced off my head. *Ouch!* My head was throbbing already, having connected with the boom in an accidental gybe while I had been distracted by my friends' antics. It had split the skin and raised a nasty lump, but I was determined to continue.

'Bowen or bust,' said Rex with an evil grin.

After clearing Broken Bay, *Mistral* fled northwards up the coast. We took turns clinging to the tiller, solid sheets of spray flying across the decks. Sundown made me gloomy, as the distant land merged into the darkness, clouds obscured the stars, the sea turned into a pit of slimy monsters, and the long night ahead, as we crossed the notorious Stockton Bight, seemed filled with menace. I had a severe headache, to add to my misery.

I had to laugh, though, when Rex appeared in the hatchway. The poor boy was feeling seasick, and he vomited listlessly into the cockpit. His face really was green, and he looked like one of those sad puppets in a Punch and Judy show. The driving spray, which was being flung right over the boat, soon washed away his bile.

Despite seasickness, however, and this being his first night at sea, he coped magnificently when it was his turn to steer. He confidently took the tiller and held the boat on course. This was not without its challenges, as the analog compass card swung wildly from side to side. After watching him for a while, I tied him on with a bowline, since we had no harnesses, and strictly cautioned him not to undo it under any circumstances, then slipped into the cabin for a snooze.

On his next watch, he insisted on tying the knot himself. Not only had he managed to hold the boat on course, he'd untied the bowline, worked out how to do it, and retied it, several times, until he could tie it in the dark, like any old salt. I realised that this kid was special.

The township of Nelson Bay, in Port Stephens, seemed like heaven after our wild night at sea. It was such a pleasure to stroll up the street, eat an ice-cream, and sit quietly on a park bench, soaking up the winter sunshine.

*At the public wharf in Nelson Bay, Port Stephens. There is a artificial harbour and expensive marina here now.*

After a long, sweet sleep, I went to see a doctor about the painful wound on my head. He was unimpressed when told the injury had happened more than 24 hours beforehand, and I was unable to seek medical advice earlier because we'd spent the night at sea.

Thinking I was a deckhand on one of the local fishing boats, he growled, 'Skippers like yours should be locked up.' Rex and I had a good laugh about that. However, there was no concussion, just a deep, infected gash.

Back on the boat, we discussed our next move, agreeing to leave in two days, assuming my head had improved. In the meantime, Rex amazed me by taking everything out of the boat and washing it, even our blankets. Then he scrubbed out the interior. Everything sparkled. I didn't wash my blankets from one year to the next.

We sailed for Coffs Harbour two days later. My head was on the mend, and the wind, which had switched to the south, was forecast to moderate. Out in the ocean, there was a significant swell rolling up from the SE, but at least it was sunny.

*Rex steering a careful compass course as we head north from Port Stephens, something that was vitally important in the days of physical navigation on the coast or near offshore reefs. We had no self-steering mechanism.*

The wind seemed to be freshening, rather than easing as forecast, so we quickly tucked two reefs in the mainsail. After a bit of a struggle on the minuscule, lively foredeck, I managed to pole the jib out on the opposite side. At this stage, when deck work was required, I got Rex to steer while I adjusted the sails, but it wasn't long before he worked out what to do.

*Sailing fast to the north with a touching faith in the weather forecast. The lines tied to the rigging are there for extra security when going forward.*

Then, with a touching faith in the weather forecast, we surfed happily northwards, confident that things would ease by nightfall. Our sole weather information came from listening to the ABC Farmers' Report on our little transistor radio. It was usually accurate, but invariably late. Whatever they predicted, we'd already had it.

With the wind behind us, *Mistral* began to really eat up the miles, surfing for prolonged periods. In the late afternoon, menacing black clouds rose astern. The sea became a deep and gloomy green, overlaid with brilliant white crests flashing in the last of the sunshine. *Mistral* was now surfing almost continuously in the rising wind, which had a hard edge to it, as if it could bruise you with just a little extra puff. Our wake was extraordinary. It flew out behind us as if we were being propelled by a turbo-charged diesel engine.

While we were debating what to do, *Mistral* suddenly passed over some sort of fish trap, with a long, thick rope just below the surface, attached to a row of buoys. In the distance, lost among the waves, were the warning flags. That rope could have tripped our keel, or ripped the rudder right off, which was transom-mounted, without a skeg to support it.

*By late afternoon, Mistral was surfing almost continuously, as menacing, black clouds rose astern.*

We decided to slow down immediately. With some trepidation, I slithered onto the foredeck and took down the jib, sitting on it while I unclipped it, then slithered back to the cockpit, trying to hang onto the boat and the sail at the same time. Then I drew a deep breath, grabbed the storm jib and slithered forwards again.

Unfortunately, having failed to familiarise ourselves with the storm sails, things didn't go according to plan. I'd grabbed the spitfire jib by mistake, which was a lot smaller than the storm jib. It was supposed to be our hurricane sail. Rex had joked that he was going to put it in a shoe box labelled: *for ultimate storms only*, then post it to his mum. We did not expect to need it on this trip. It was too small to balance the reefed mainsail, and, worse, I had hoisted it upside down. However, it set reasonably OK like that, so I left it and slithered back to the cockpit. In those conditions, the foredeck was no place to linger. It was like dancing on a window ledge.

Suddenly *Mistral* broached. Sitting to leeward, I was up to my armpits in water, and it felt as if the boat was falling on top of me, but Rex, scrambling up onto the windward rail, didn't even get his socks wet, the piker. Within seconds, the boat was back upright, running off downwind. Hardly any water entered the cabin, despite the top companionway board being stowed below. It was soon located and shoved in. Then I leapt up and dragged the mainsail down in record time. *Mistral* continued to thunder along under the spitfire jib, which was less than a square metre in area. Our respect for the boat soared.

At dusk, the wind was up to 40 knots, we discovered later. The boat hurtled down the waves, exceeding 10 knots for long periods at times. Without a doubt, we should have streamed our drogue, but although the tyre and chain were lashed on the stern rail, the storm warp was still in the aft locker, tangled up with all sorts of stuff, and there was no way we were going to open the hatch in those conditions. All we could do was hang on and hope. There were times during the night when the boat was surfing so fast it started bouncing down the face of the waves.

*Mistral takes off down the face of a wave. The needle of the speedometer was regularly hitting the 10-knot stopper.*

The wind eased shortly after dawn, and late that night we entered Coffs Harbour, having sailed 193 miles in 36 hours. The stillness and sense of security within the inner harbour were exquisite. It was well past midnight, and there was nobody else around.

After mooring fore and aft to a set of piles, we sat in the cockpit, looking entranced at the reflections mirrored on the flat water, feasting on a tin of French Onion soup and stale bread. It was our first hot food since Port Stephens, and no king ever had such a banquet. Then we inserted our bone-weary bodies into the narrow quarter-berths and slept like drunks.

We stayed in the marina for three days. Rex, once again, took everything out of the cabin, washed and dried it, and scrubbed the lockers. I was beginning to feel distinctly grubby in his presence.

*Rex walking to the summit of Muttonbird Island, Coffs Harbour.*

We visited Rex's mum, went out to dinner, and walked to the top of Muttonbird Island, which forms part of the harbour enclosure. Ever since that visit, I've retained a deep fondness for Coffs Harbour. My delight in being there was undoubtedly heightened by contrasting this snug harbour with our hard passage north from Port Stephens, but I think it also had something to do with the fact that I was happy, despite the cold, and the discomforts of sailing *Mistral*. Joy transforms everything.

We sailed for Mooloolaba one sunny morning, drifting along over calm seas under clear skies, thoroughly enjoying the easy going after our earlier passages. We were becalmed for a while off the reefs that lie near Evans Head, enjoying a spectacular sunset as compensation.

*Becalmed for a while off Evans Head.*

*Easy sailing north of Evans Head, running before a steady 15-20 knot SE wind.*

24 hours later, a swell from the south presaged the arrival of a 15-20-knot SE wind, before which we happily surfed under full mainsail and working jib poled out, wing and

wing. Once again, we were gobbling up the miles, though this time the weather remained moderate, and by now we were in tune with the boat, the sea, and each other.

The days were sunny, although still cold, and the nights were a symphony of stars. In the easier conditions, Rex and I relaxed, amusing each other with an endless series of jokes and stories. I had the feeling that I could have sailed on forever with Rex on *Mistral*. So much for thinking *Adrienne Matzenik* was too small, although *Mistral* was a later model with an enlarged coachroof. It was a beautiful passage.

The day after we arrived in Mooloolaba, we were walking up the road from the yacht club when a van screeched to a halt. To my surprise and delight, Daniel jumped out. He and Nicole had just returned from France and purchased a new boat, *Aito*, for a return to the islands. Nicole, as usual, had already started customising the boat, using her considerable craftsmanship, to change it from a racing boat into a family cruiser.

Our intention had been to cross Wide Bay Bar after leaving Mooloolaba, and sail up the inside of Fraser Island through the Sandy Straits, but you need calm weather to make that crossing, and the SE winds were still blowing fresh. It was also a far more dangerous place in those days than it is today, as finding the channel across the bar depended on sighting the leads, a challenge since the bar is well offshore and the leads often obscured by haze. It was imperative to have the sun behind you, a clear horizon, a swell of less than 1.5m, a rising tide, and clear visibility. Today, although the bar still requires moderate weather, you can get GPS waypoints from Tin Can Bay Coastguard to plot your course through.

After four days, Rex was anxious to get going, as he was expected back at work in Bowen by the end of July. We decided to sail around the outside of Fraser Island, past Breaksea Spit. This can be a notoriously hard passage, and my heart was pounding as we sailed out of Mooloolaba's breakwaters.

*Mistral* reached Breaksea Spit on the second night out, beating against 25-knot NE winds. Rain squalls blotted out Sandy Cape lighthouse for substantial periods of time, leaving us alone in the dark. The tide was ebbing to the NE, which set us away from the reef, but it was in opposition to the wind, waves and ocean current, creating severe overfalls. *Mistral* was leaping off waves and crashing into the troughs as if falling on concrete.

I had no time for sleep. When not steering or looking for damage, I tried to update our position, kneeling on the cabin sole and plotting a dead-reckoning position on the chart in front of me. We had no chart table, unfortunately, as *Mistral* was deemed too small by the designer. Mind you, both *Sopranino* and *Trekka* had proper chart tables. It is a question of priorities.

Occasionally I was able to get a bearing on the intermittently visible lighthouse, but it was impossible to get a fix, and even the bearings had to be treated with some caution due to the wildly swinging compass card. By this time, I had learned to average out its swings, rather than try for a single, accurate reading.

In those conditions, dead-reckoning navigation could most accurately be described as educated guesswork. *What was the combined effect of the northeast-setting ebb tide, the south-setting East Coast Current, and the boat's leeway?* Old-timers in Mooloolaba told me to plot a course at least 15 miles off the lightship at the northern end of the spit, to give myself sufficient margin for error.

The boat fell off waves so heavily at times that I was convinced I'd got it all wrong and we were on the bricks. Then *Mistral* would leap up again and fling itself at the next wall of water. The seas were only two metres high, or so, but they were very steep. We were afraid that the pounding would split the hull, and I regularly shone the torch around looking for cracks. There were none, but later we found a little water in many of the lockers. We had no idea where it came from, since under normal circumstances, *Mistral* did not leak.

Trying to hold onto the torch, dividers, parallel ruler, pencil and chart, let alone think, was a rigorous task. I needed to decide at what point it was safe to bear away westwards for the mainland. Too soon and we would end our days on Breaksea Spit; too late and we risked being set among the reefs further north. Eventually I was sure we'd covered the distance, and the lights of an outward-bound ship inshore of us, obviously also rounding the Spit, confirmed it. Thankfully, we bore away, putting the wind and waves behind us, and bringing the incessant pounding to an end.

Sometime later, bearings on the intermittently visible lighthouse confirmed we were past the shallows and could relax. Our strategy had worked, but looking back at passages like this from the GPS age seems astonishing, even to those of us who lived through them.

There was still a long way to go to Gladstone, however, and we were both desperately tired. I had not had any sleep for about 15 hours, and was also mentally exhausted. Rex, that stalwart, despite his own tiredness, offered to take one of my watches the next morning, meaning he steered without a break for nine hours.

Despite his long stint at the helm, Rex was not keen to go below when I resurfaced. Once again, the boat was squirting along as if there was a jet engine in the bilges, with the main vanged out one side and the jib poled out the other. Water roared out behind the transom, leaving a broad wake. Rex clambered up onto the cabin top and sat there grinning. 'How often do you get a sail like this?' he asked rhetorically. He clung to the rigging, wet hair plastered across his face, looking back at me on the tiller. 'You are the toughest guy I've ever met,' he added, but I knew it wasn't remotely true.

By late afternoon the sky had cleared, and I identified the hills behind Pancake Creek, matching their profile to one drawn on graph paper, taken from the contours of the chart. Harry had shown me how to do this, and I was proud of myself. We stood on through the dusk, making our now customary night entry into Gladstone Harbour.

*Moored to the piles in Auckland Creek, Gladstone, we took everything out of the boat, washed and dried it.*

Most of our gear, clothing and bedding was damp and salty after this passage, so we once again took everything out of the boat, washed and dried it, then did the same to the interior before putting it all back in. In those days before the marina was built, visiting yachts made good use of the Gladstone Yacht Club, and we were no exception. No matter how tough you are, and neither Rex nor myself claimed to be, there was nothing like a hot shower to wash away the salt, followed by a cold beer and a juicy steak when we came ashore. I still like hot showers, but will pass on the beer and steaks.

It was warm now, just south of the Tropic of Capricorn. The sun shone nearly every day, and the hardships lay behind us. We took *Mistral* up through the Narrows inside Curtis Island, where Guzzwell had taken *Trekka*. It was like a pilgrimage for me.

The first night we anchored at Laird Point, just inside Graham Creek. It was pitch black, and I was sitting in the cockpit, peeling potatoes, as happy as Larry, whoever he was, when a dreadful, slurping noise started up behind me. Envisioning huge crocodiles, I leapt into the cabin in alarm. Rex laughed. 'It's just a dugong, stupid,' he said. 'They don't bite.' But nothing could induce me to go outside again.

*Fogbound on the tropic of Capricorn! Rex gives up on the recalcitrant outboard motor.*

The next morning, we woke to thick fog, just miles south of the Tropic of Capricorn. I have since discovered that it is quite common in winter there. We also had trouble with the two-stroke outboard motor. Some idiot in Gladstone had put too much oil in the petrol (we both blamed the petrol attendant) and it clogged up the spark plugs.

Rex tried for some time to get it started, getting redder and redder in the face and swearing profusely, but without success. I stood on the foredeck taking photos. It was the first time I'd witnessed one of the famous tantrums his brother, Tom, had warned me about. 'Such language,' I joked, 'out of the mouth of sweet-faced children.' Suddenly, he threw down the tools and walked forward. I thought he was going to come up onto the foredeck and hit me, but instead he went below and dropped tearfully onto his bunk.

I tiptoed aft and sat quietly in the cockpit, peering cautiously down at him. He had his back to me, staring at the bulkhead in frustration. After a suitable interval, I took the spark plugs out and cleaned them in kerosene. The motor burst into life first pull of the starter-chord. With a broad grin, Rex bounded out of the cabin like a puppy, and we were on our way.

The fog, and the recalcitrant motor, had delayed our transit of the shallows, known as the cattle crossing, however, and we ran aground on a falling tide. At low tide, *Mistral* was completely out of the water.

*Mistral aground in the Narrows, at the northern end of Gladstone Harbour. Rex climbs around the deck to burn off excess energy. 17-year-olds! I lay in the inflatable dinghy, watching him.*

In the distance, Mount Larcom's jagged profile speared the tropical sky. Rex walked around scrubbing the bottom of the boat and exploring the mangroves, while I lay in the inflatable dinghy enjoying the sunshine. Neither of us thought about crocodiles. They were less numerous back then than they are now, decades after the species were protected by legislation, but they might have been there.

There were other, invisible teeth on the prowl, unfortunately. I was bitten badly by sandflies and it took several days for the bites to heal. It was my first experience with this scourge of the Queensland coast. If there are mangroves nearby, and a rising tide, beware. We had to spend the night there, drying out again in the early hours of the morning.

Rex, still being something of a callow youth, also poached a mud crab from a crab pot up a side creek, but just as he started cooking it, the crabbers arrived to clear their pots, and came over to check us out. Rex hastily closed the hatch and sat on it, hoping they did not notice the distinctive smell of boiling crab. These folk, who carried rifles to protect themselves, supposedly, from crocodiles, were rumoured to take pot-shots at poachers.

*Rex taking the opportunity to give the botom a clean.*

The next day we beat out from the under lee of Curtis Island, then had a lazy sail across to the Keppel Islands. It is only about 15 miles, but we took several hours to arrive, lounging happily in the sunshine.

This was the life, sailing across a bejewelled sea, towards enticing, green, tropical islands embroidered on the azure horizon, palm trees hanging languidly over shimmering, yellow beaches, ripe coconuts lying on the hot sand. Actually, I have never seen a coconut tree on Great Keppel Island, but you get the idea.

Despite the temptation to continue north before the fresh trade wind that scurried us along after leaving the Keppels, we detoured into Rosslyn Bay, as we were on a mission, hitchhiking into Yeppoon in quest of one last feed before tackling the empty islands to the north. One gets sick of instant noodles occasionally. After a few beers in the pub, we ordered a Fisherman's Basket each, which came overloaded with every form of seafood imaginable, straight from the local trawlers. It was so scrumptious, we ordered second helpings.

This made the chef come out to look at the pigs, because our request was unheard of. Usually, people asked for doggy bags to take the surplus home, and he took pride in his generosity. Our second helpings were more humongous than the first, as if he was determined to defeat us, but we stuffed it all in, knowing what lay ahead. It was 15 miles back to the boat but we couldn't get a lift. Perhaps the locals had heard about us and were afraid we'd eat the upholstery, so we had to waddle painfully all the way.

*A slow drift to Port Clinton, north of the Keppel Islands. The foredeck was no longer a scary place.*

After a slow drift to Port Clinton, during which we exhausted our petrol supply, the next day dawned wet and windy. Rex, however, was getting anxious about his commitments, so off we went. Soon we were swooping along, surfing in the gusts, soaked to the skin in the squalls. It was too warm to wear wet-weather gear, which in those days was clammy at the best of times. Besides, after the hard going down south, this was just a frolic; one-metre following waves, warm winds, and only 20 miles to the next anchorage. Two large ketches passed us, going south, pitching painfully into the steep waves.

'Mugs,' said Rex, laughing.

We anchored behind Cape Townsend with plenty of daylight to spare. The wind was *really* howling now, and the rain pelted down, but we were nicely sheltered, with only a slight surge working into the bay.

A little later, a large yacht came in and anchored further out. I recognised it as the 74', carvel-planked Alden ketch, *Enid*, last seen in Rushcutters Bay in 1973. Launched in 1962, it proved to be seriously fast, holding several race records. Since we crossed wakes in Cape Townsend, *Enid* spent many years in the Whitsunday Charter trade. After being damaged by Cyclone Debbie in April 2017, it was fully restored and was entered in the 2023 Sydney Hobart Race.

The next day, the rain, and *Enid*, had gone, but the wind was still howling. Off to the north, the horizon looked like crocodile teeth, so we decided to stay at anchor, lazing in the cockpit, reading, fishing unsuccessfully and dozing. Rex went ashore to do some beachcombing, and I lay in the cockpit, feeling warm and dreamy. This area is a military training zone, and the place has a remote and lonely air. There are no settlements ashore, and civilians are not permitted above the high-tide mark.

The following day we sailed to Middle Percy Island. When *Mistral* arrived at West Bay, there was a fresh westerly wind blowing straight into the fabled anchorage, and a nasty swell rolled around from the east, causing the boat to bob and rock around like a football fan. I had been building this place up to Rex for weeks, and we were both disappointed. We decided to get underway after supper. There was no point staying if we couldn't sleep.

We had a barbecue ashore that evening, behind the A-frame shed, with another young guy, Geoff, who was sailing a 12' open dinghy alone from Brisbane to Cairns. He made our voyage look decadent. He was a trained shipwright, not long out of his apprenticeship, and had built his dinghy, customising it with watertight storage lockers, making it unsinkable in case it capsized.

He had cut his charts up into strips and laminated them. The chart of the day was folded, concertina-style, and kept in the top pocket of his shirt, so he could pull it out and check it when needed. He'd written all his courses, bearings and passage notes on the chart, as he was unable to leave the tiller to navigate. The system was obviously working, as Middle Percy Island is 50 miles off the coast, although there are a number of islands along the way to orient oneself.

As we sat in front of the fire, eating our simple meal, which was probably instant noodles again, or maybe a real treat like tinned sausages, I looked out across the bay. The palm fronds, clicking in the breeze, were black silhouettes against the sunset sky, and the sea muttered on the beach. It all seemed so far away from Sydney, and the desire I felt to get out of that place permanently was almost overwhelming. If only I hadn't sold *Adrienne Mutzenik*. I was both thrilled to be here with Rex and sad that I'd squandered my chance to escape.

By the time we went back on board, the tide was out, and *Mistral* was almost in the surf. Rex dropped the eggs we had bought from the honesty-store in the A-frame, was violently seasick, and then collapsed into his bunk, so plans for an early departure were abandoned. Not feeling too good myself, I threw the spare anchor out as far as I could and hoped for the best.

If we'd had any petrol left, I would have re-anchored further out, but *Mistral* was strictly a sailing boat now, at the whim of the wind, and I didn't fancy trying to sail off the anchor unless I had to. I should have rowed another anchor out, but didn't feel up to it on my own. In later years, I'd learn to do what had to be done.

We always double-reefed the mainsail and hanked on the storm jib before nightfall in these open-roadsteads, in case we had to beat out against an onshore gale at short notice, so we were ready to go if the weather deteriorated. I kept an anchor watch instead, until the wind died out in the early hours, and then crept sleepily into my bunk.

There was little wind the next day. Unable to reach Scawfell Island, 60 miles north of Middle Percy, before dark, we rode the tidal stream over to the neighbouring islands of St Bees and Keswick, anchoring on their northern side at midnight. I was disappointed to miss Scawfell, of which I had such fond memories from 1977.

The following night, utterly becalmed, we were forced to anchor in the open sea, just south of Thomas Island, to stop going backwards with the tide. The ocean was flat, but it was eerie to be anchored in the middle of nowhere. To add to the atmosphere, we listened to a classic black comedy on ABC radio, *Arsenic and Old Lace*, about some old ladies who poison their tenants. When the play was finished, we sat in the cockpit and competed to tell each other the scariest stories we could remember. There was no moon and the night was pitch-black. We were like two big kids on a sleep-out, and ended up in helpless giggles.

The next day there was no wind again, and we were running out of food. Rex cooked the last of the rice and we ate it with mustard. *Delicious.* Later we cut the last lemon in half and sucked it disconsolately, as we drifted this way and that. In the distance, other yachts motored happily along. It was painful, imagining the feasts *their* crews were having. We began to chant, 'Food, Food, Food,' every time we saw another boat, like monks chanting religious mantras. The other yachts were too far away to hear, but it was amusing and distracted us from our hunger. Alas, there was no pie in the sky for us.

*It was so calm that we launched the dinghy and took photos of Mistral drifting along.*

As we drifted close to the southern point of Dent Island, near the lighthouse, my old practical joker streak rose up. I've always loved practical jokes, and was notorious for it as a boy; it often got me into trouble. This time, I imitated slack-jacked incredulity while peering under the boom.

*Would you believe anything this man said?*

Rex, who was in the cabin, jumped up on deck, saying, 'What is it?'

'An elephant,' I shrieked, pointing at the island.

'Where?' he said, looking under the boom at the forested hillside, only to realise that my shrieks had changed from incredulity to amusement.

Such are the things that boredom drives you to. I tried explaining this, but it was a long time before Rex forgave me. He spent our remaining time together trying to exact revenge. 'Look, a killer whale,' he'd shout, but he never fooled me.

Late that night, we had to paddle with the dinghy oars to escape the clutches of a small ship intent on running us down. Of course, it couldn't see us, because our battery was flat and we had no lights, not even a kerosene lamp or torch. Everything was running on empty.

We finally managed to anchor off the resort on South Molle Island shortly before dawn. In 1977, on *Silver Heels*, I had to take my trousers off on the beach here, to swim after a wayward dinghy (thinking about large tiger sharks). Upon returning in wet underpants, I discovered three young women from the resort had confiscated my clothes. They made me beg. Perhaps this story was the reason Rex was so keen to go ashore that morning, or it might have been mere hunger.

We had no money though, and I was too proud to beg for food (or was I scared the girls were still there), so I persuaded Rex to push on to Airlie Beach. Luckily, a nice breeze sprang up as the morning progressed, and we made it without too much effort, despite threats of cannibalism from the disaffected tomato picker. Once ashore, after visiting a bank, the gluttons emerged from our ascetic souls, and we demolished an enormous pile of food. By the time we had finished, we both felt sick.

When we had recovered from indigestion, we sailed before a brisk wind to Woodwark Bay, where our keel bounced on something hard in the middle of the night. Many years later, I discovered an uncharted coral bommie in this anchorage, most certainly the culprit. It is invisible at high tide.

The next day, in cloud-dappled sunshine and a moderate wind, we sailed through Gloucester Passage and into Bowen, five weeks after leaving the Hawkesbury River.

On that last day of the voyage, the majestic green peaks of Gloucester Island were etched against a deep-blue sky, and the sheen of *Mistral's* orange topsides cast reflections onto the waters of Edgecumbe Bay, which glowed with that aquamarine hue peculiar to shallow tropical waters. I was happy to have made it, to see Rex shining like a lighthouse in his pride and joy, but I was sad that it was coming to an end.

I only spent a week in Bowen, although Rex wanted me to stay for the winter and get a job picking tomatoes with him. I really wanted to, but felt I had to return to Sydney and keep working towards a boat of my own. Perhaps I should have abandoned everything in Sydney and stayed, because there really wasn't much back there for me.

# Chapter Fifteen

# A Schooner in the Offing

*There's a schooner in the offing,*
*With its topsails shot with fire,*
*And my heart has gone aboard her*
*For the Islands of Desire.*

Richard Hovey

***

It was still mid-winter in Sydney. The day I arrived back was a typical bone-crusher, foggy and damp, with a bitter wind off the snowfields that struck you like the business end of a cricket bat. As my train rattled its way through the cluttered inner-city suburbs, I looked out at the grime-encrusted, graffiti-emblazoned walls, at the garbage strewn alongside the railway tracks, at factory chimneys spewing into the winter sky, and ached for the warmth, natural beauty, and promise of the islands, not to mention my easy friendship with Rex.

To overcome my ennui, I decided to go over to Dangar Island. David Lewis and Yvonne Liechti were in Indonesia, researching traditional navigation systems, but they'd invited me to spend time in their house, and had also asked me to look out for an old friend of theirs, Kurt Ashford, who was due in from Fiji on his gaff-rigged schooner, *Ishmael.*

Highly regarded as a traditional shipwright and rigger, Kurt had roamed the Pacific Ocean for most of his adult life on home-built schooners, from Seattle to Auckland. He had spent significant periods of time in Tahiti, Samoa and Hawaii, and spoke fluent Tahitian. He had re-rigged the square-rigger, *Falls of Clyde*, in Honolulu, and *Ishmael's* masts were sourced from timbers shipped into Hawaii from British Columbia for that project. They'd been intended for the ships' yards but proved to be too small in diameter.

David Lewis first met Kurt on the island of Raiatea, in the Society Islands, in 1965, where Kurt had been rebuilding his schooner, *Sea Gypsy*, after it had been swept over the barrier reef in a cyclone. Not long after that meeting, Kurt sailed *Sea Gypsy* to Honolulu, where he built *Ishmael* over a period of seven years. This was his first long voyage in the new schooner. He had a legendary reputation among ocean voyagers; even David was in awe of him. He paid no heed to shore-side conventions, didn't own a pair of shoes, and usually went shirtless, relying on his skills, wits, and charm to survive.

Wherever he went, people begged him to work on their boats, and his friends were fiercely loyal. One Tahitian man, Metata, having sailed on *Sea Gypsy* for several years, grew vegetables on Bora Bora and gave them to passing yachts in honour of his old skipper. Croix had mentioned Metata in his letter from Bora Bora, when *Iorana* passed through in 1971. Kurt sounded like the ultimate deep-sea vagabond and I was longing to meet him.

To get to Dangar Island, I took the local ferry, the *Sun,* from Brooklyn public wharf. The cold, damp fog was so dense that nothing was visible a few feet beyond the dock. Like a wet blanket, it seemed to muffle all sound, other than the throb of the diesel, and the swish of the bow wave. The skipper, who knew the tidal streams well, steered a careful compass course, and I stared into the void. The fog brought back memories of the Narrows in *Mistral.* I grinned wryly, remembering Rex's hilarious tantrum when the outboard motor wouldn't start, and in that moment of memory, hope flickered.

I was still smiling when a huge, gaff-rigged schooner loomed up out of the fog, anchored off the eastern side of Dangar Island. It looked like some vision from the 19th century, painted in traditional American schooner colours, with a blackish green hull and buff deckhouse. As I discovered later, the jib-boom, atop the bowsprit, extended 22' beyond the bows, and the mainsail boom extended several feet aft of the transom. The ship had tall, raking masts, with ochre-coloured, tan-barked canvas sails furled loosely on the spars. I only had a brief glimpse, then the outline of the schooner disappeared into the murk just as quickly, like a phantom ship.

Once ashore on the island, I scuttled around to David and Yvonne's, quickly untied their commuter launch, and motored out to the schooner. There was nobody aboard, but I was sure it could only be Kurt. Sure enough, passing close astern, I saw the name, *Ishmael*, and beneath it the port of registry, *Honolulu*, carved into the high, galleon-like transom. Drifting in the schooner's lee, I caught a whiff of Stockholm tar, wet pine decks, wood oil, and the patina of salt that lay over everything. There is nothing like the smell of a traditional sailing ship, it is a fragrance that walks straight off the pages of Stevenson and Melville. I thought *Ishmael* was the most romantic vessel I had ever seen.

*Ishmael anchored off Dangar Island,*
*Hawkesbury River.*

Kurt built *Ishmael* from a half model he had carved. When people asked who designed the schooner, he said, 'Nobody.' If pressed, he said it was built the way his granddaddy used to build schooners, by eye. Envisioned as a trading schooner, *Ishmael* had a large cargo hatch amidships, relatively shallow 6' draft, internal ballast, and tandem centre-boards. The hull was triple-planked, with two watertight bulkheads, one between the forepeak and cargo hold, and the other between the hold and engine-room. There was an extra layer of planking, set in butyl mastic, below the waterline, a traditional form of sheathing against teredo worm.

There was no cockpit. The helmsman stood on deck alongside a 4' diameter wheel, mounted aft of a box that housed the steering chains. The chains passed down through hawsepipes on each quarter and were attached to the trailing edge of the rudder, as in old sailing ships. There was a tin can filled with used engine oil wedged inside the base of the steering box, and each watch slopped a bit on the chains to lubricate them. There was an old hessian sack on the deck, to absorb oil drips, but the oil still tended to track below. It got into your bunk, and even your hair, eventually, as I was to discover.

Numb now with cold, I took the launch ashore and went up to the house, to see if Kurt was there, although there was no telltale dinghy tied to the wharf. I learned later that it was on the nearby beach. Opening the back door, which was never locked (you could do that on Dangar Island), I went into the living room and called out, 'Anybody here?'

A thin, middle-aged man appeared from the bedroom, pulling on his trousers. He had an unkempt appearance, with a scruffy black beard, ragged clothes, and a bad limp. David had mentioned Kurt's leg. He took a bullet in it as a 19-year-old soldier in the Korean War, but I was still shocked by how wasted it looked. He had been left behind in the field, spending days hiding from Chinese soldiers before being rescued. The wound on his shin had never healed, despite several skin grafts, and he was in constant pain.

'It's so bloody cold, we've been hiding in David's bed since we arrived,' Kurt said. 'This is Ginna Cizek, my crew.' He gestured towards a young, auburn-haired woman who followed him out of the bedroom, shyly buttoning up her clothes. Ginna seemed to possess an interesting blend of shyness and self-confidence. She looked like a tomboy, with strong, practical hands, and proved to be adept at physical tasks, but there was something delicate about her at the same time, an innate beauty. Like Kurt, she had a distinctive aura.

'Shit,' Kurt added, 'where does David hide his booze? I've ransacked this place.'

'I don't think he drinks much,' I replied.

'Drinks enough of my bloody rum when he gets the chance,' Kurt shot back, with what I came to recognise as his trademark dry humour.

I should add here that David was never a big drinker. He might have had a couple of extra rums on special occasions, but I never saw him drunk. He was too focused, too disciplined. When I was staying with him, he'd usually have one glass of red wine with dinner, then refill his glass afterwards and retire to his study, to continue writing or researching.

Well,' I said, pulling a bottle out of my bag, 'it just so happens that I've got one. David said you might be here when I got back, and he told me you liked rum.' We got down to the serious task of warming our bellies with booze and telling yarns, as sailors do. The sky was almost dark by the time we came to the end of the bottle. Ginna had already slipped away for a shower and another snooze. Kurt had a funny look on his face.

'I suppose you're wondering if I really did murder my wife,' he said. I stuttered a denial. The thought *had* crossed my mind, though I'd immediately dismissed it. David had told me that Kurt's wife was murdered shortly after *Ishmael's* launching, and how Kurt had initially been charged with the crime, largely due to his unorthodox lifestyle. Charges were later dropped because the police had no evidence to implicate him, but by then his life had spun out of control. *Ishmael* had been built to voyage around the Pacific with his family, carrying the tools of his trade and occasional cargoes, but he said that everything seemed futile now. He was drinking heavily, separated from his children, and thinking of selling the schooner.

'It's OK,' Kurt said, though it clearly wasn't. 'Everybody wonders. That's the problem with this sort of thing; some mud always sticks. Things are never the same.' He looked unbearably sad. I wanted to give him a hug, but he didn't look like the sort of guy who could handle that. I reached over and gave him a friendly nudge with my fist, saying that David had assured me he was innocent. Kurt grinned raffishly. 'That bastard. When's he due back?'

David and Yvonne duly returned and the party continued. I spent as much time as possible on *Ishmael*, sailing around the local waterways, neglecting my half-built ferro-cement boat. I occasionally did some desultory work on the hull, but my heart wasn't in it, never had been. I drank more rum than I'd ever done, frequently getting dizzy spells and passing out on *Ishmael's* deck.

Kurt was always joking about that. 'You're the most hopeless drunk I've ever met,' he laughed. 'Here, have another rum, it'll make you feel better. Feed the hand that bites you.'

'I'd like to sail with you back to America,' I said one day, looking at Kurt wistfully, 'but I've got no money.'

'I'd like to take you,' he replied. 'You have a real feeling for the old ways, and you're one of the best helmsmen I've ever sailed with, but I don't have any money either. You'd better stick to what you're doing. Finish that boat of yours and meet me in the islands.' I think we both knew it would never happen.

*David Lewis, left, and Kurt Ashford, on the aft deck of Ishmael, enjoying a glass of rum.*

*Ishmael* became the stage for many boisterous social events, from parties on deck to sailing around the river with a large gang of would-be pirates. Once, we sailed through the narrow, small-craft channel at Whistling Kite Point, on the south side of Dangar Island, just inches off the rocks. Thank goodness there were no commuter boats coming the other way, or if there were, they must have scattered.

On another occasion, I was sailing *Ice Bird* and *Ishmael* was chasing me, just off Croppy Point, near Little Wobby. *Ishmael's* crew, including numerous Dangar Island friends like Ann Howard and her boys, Paul, Lincoln and Erle, were leaning over the leeward rail as they closed in, towering over us, yelling like cutthroats and throwing flour bombs. I was trying to throw buckets of water back, but they had the upper hand, especially after I lost the galvanised bucket overboard.

*Sailing around the Hawkesbury River on a cold, winter day. I am on the wheel and Kurt stands beside me in khaki overalls. Photo: Scott Mitchinson.*

It was a sad day, in February 1980, when the schooner sailed out of the lives of Dangar Islanders, bound for Sydney Harbour to gain outbound Customs clearance for New Zealand, and then back into the Pacific Islands. I sailed aboard down to Sydney Harbour, where we anchored in Watson's Bay. In those days, before the death of romance, it was the Customs clearance anchorage for sailing vessels, as it had been since the time of square-riggers.

*Here, on a quiet day off Dangar Island, Ishmael's canvas sails are hoisted to dry in the time-honoured tradition.*

Everywhere that *Ishmael* went, the boat created a sensation, and it was no different in Watsons Bay. When we launched our 18', black-hulled Whitehall skiff, and rowed ashore like 18th century explorers, the proprietors of Doyle's, the famous seafood restaurant, invited us to eat at their establishment for free, for as long as the schooner lay anchored off. They did not even ask Kurt to wear shoes. I had not eaten so well in more years than I could remember. So, I thought, this is how the rich and famous live. It was only fish and chips, really (we were too shy to order lobster), but there is no doubt that the spectacular view added to the allure.

*The 18' Whitehall skiff was secured upside down on these bearers at sea. Note that the foresail gaff overlaps the mainmast when lowered, and rests in a timber chock on the deckhouse roof. The clew of the foresail also overlapped the mainmast, something that Kurt said was vital to windward performance in a gaff-rigged schooner.*

Eventually, with a heavy heart, I cadged a ride ashore on the Customs launch after *Ishmael* had been given outbound clearance. I doubt you could do that today, now that Customs have been renamed the Australian Border Force, and stern-faced officers are decked out in black, quasi-military uniforms, bristling with weapons. The officers who gave me a lift ashore were full of smiles, and joked that they hoped I wasn't smuggling contraband, because Ginna had given me a beautifully-carved wooden box as a farewell present. I don't think the borders were less secure back then.

I went up onto South Head to watch *Ishmael* sail away. Sitting alone on the cliffs, I watched the schooner poke out into the Pacific Ocean, that long bowsprit rising and falling majestically as the ship breasted the SE swell. It was hard to believe that I would never again lie on those warm, pine decks, inhale the scent of Stockholm tar and tanbarked canvas, trace the intricate web of the topmasts against the stars, drink rum and share stories with Kurt and Ginna. I watched the rust-red sails, etched against an immensity of sea and sky, until they were mere dots that blurred and vanished into the purple haze of the horizon, by which time my eyes were also blurred.

The days turned into weeks without news of *Ishmael*, and the memory of those marvellous days began to fade. The only thing to do was throw myself back into my boat, finish it and sail away. The lure of the deep-blue sea had never been stronger. I found a job, bought some timber, and began building the decks. Signs of progress soothed me, and I began to dream of the day I might launch my ship. I was dreaming of sailing north and reuniting with Rex. Then Kurt's letter arrived.

*I'm sorry you haven't heard from us in such a long time, but Ginna usually writes and now she's gone and left, and I'm out of practice. You could even say I was illiterate, though I'm not sure I know how to spell that word. Ishmael is in Nelson, on the South Island of New Zealand. It's a good place and I'd be happy to stay a while, except I'm keen to get home and sort things out.*

*We had a miserable trip across the Tasman, winds SE to NE all the way, 30 to 50 knots mostly. Took us 22 days to make good 1000 miles. I feel like a miserable prick writing to ask you now if you want to join me, after not taking you when we left the Hawkesbury, especially since what I said then still stands, but I remember how much you wanted to come, and the truth is, I need a navigator. I know how to do that stuff but it just gives me the shits.*

*The way I see it, we could sail Ishmael up to Seattle, sell her, and build two new boats together. Given the little boats you prefer, you could just about build it out of my scraps. I know you hate the cold as much as I do, and it's true the summers up there are short, but we could get a big barn, put a stove in it and stay in there all winter.*

*Anyway, perhaps you could think about it and write me as soon as possible. Summer's as good as over here in Nelson, and if I'm going to get away, I'd like to do it soon, before my balls freeze, not that they're much use to me now. There aren't too many local girls lining up to take Ginna's place. Can't say I blame them. What would they want with a lame old dog like me?*

I sat on the back stairs of my cottage, clutching Kurt's letter and looking down the garden at my ferro-cement hull, but I wasn't really seeing it. A kaleidoscope of images coursed through my mind, with Kurt and *Ishmael* central to all of them. Every now and then, I focused on my current situation, the partially-built hull stark against the late-summer sky, caught a glimpse of what it could be, but then images of Kurt and *Ishmael* re-asserted themselves, pulsing like a vein in a moment of passion.

'You can have my hull,' I said to a startled friend, Scott Mitchinson, 'if I don't come back, which I probably won't.' I'd first met Scott when he came to view *Adrienne Matzenik* with a view to purchasing the boat. He decided not to, but we became friends. He also agreed to take over the rental of the cottage. Then I packed my bags, said goodbye to other friends, who were perhaps getting used to these sudden departures, and bought a plane ticket for New Zealand.

Sydney had been in the last throes of summer, but it was definitely autumn in Nelson. There had even been a dusting of snow on the hills behind town in the days before my arrival. The cold, as usual, eroded my spirits, but worst of all, Kurt was in a foul mood. He said it was because he'd expected me the night before, had gone out to the local airport and stood around in the cold for nothing, but that did not make sense. I hadn't given him an ETA, or said I was flying in from Christchurch. I took the bus.

From the moment I stepped on the boat, things were difficult. The only change I could see was that Ginna was no longer aboard. Kurt had frequently been critical of her, and it seemed that I was the new whipping boy. Whenever I opened my mouth, I seemed to put my foot in it, and, despite my desultory efforts at amateur boatbuilding, I'd never been much good with tools, especially with a critical eye on me. Every day brought new miseries. I felt useless, and my 28th birthday passed without declaration on 12 April 1980.

I tried as hard as I could. I've always hated heights, but when Kurt told me he was going to hoist me to the top of the fore-topmast, 75' above deck, because I was the lightest person on board, to grease the galvanised steel forestay, I quietly complied, even though I did not trust the gear I was being hoisted on. Once at the masthead, I had to swing out and grab the forestay, before sliding down it, wiping the wire as I went with a rag that I frequently dipped into a tin of grease tied to the bosun's chair.

Kurt had also found another crew member, 22-year-old Jim Ferris, with whom he was getting along famously. Jim was one of those quiet, competent types, of Scottish descent like many South Islanders, with a calm, handsome face beneath tidy, brown hair. He was sturdily-built, strong and courageous, and he'd already completed several projects on board.

If Jim had been a navigator, I suspect Kurt would have asked me to leave, but then Jim might have become the new whipping boy. It seems Kurt always needed someone to berate, perhaps to externalise his demons. Adding to my discomfort, my bunk was too short, and the foam mattress so thin and worn it was almost non-existent in places. And it was damp. The situation bore no comparison to those warm, carefree days on the Hawkesbury River. I'd been content to sleep on deck back then, and Kurt and I had been great friends.

It was an unhappy ship that sailed a few days later. *Ishmael* was headed, initially, for the township of French Pass, to careen the boat, scrub the bottom, and do some running repairs to the tandem centreboards, which were beginning to delaminate. The centreboards were made of steel plate, clad in plywood that had been shaped in a rudimentary fashion, then sheathed with polyester resin and chopped strand matt. It was a cheap form of construction that was guaranteed to fail sooner or later.

In this case sooner, as *Ishmael* was then only seven years old, and had only sailed for two seasons. The fibreglass sheathing had chafed against the edges of the centreboard cases, exposing the end-grain of the plywood to water-ingress. The plywood had then swelled and separated from the steel, which was now rusting. They really needed a serious rebuild but we didn't have the resources to do so.

*Ishmael was built on a tight budget, and most of the gear aboard was repurposed, like the 1948 tractor diesel engine, and the manual anchor winch seen here, which came from a decommissioned Coastguard cutter.*

On the first night out of Nelson, we were becalmed. Kurt tried to start *Grandma*, the 1948 Caterpillar tractor diesel, but could not get the pony-motor going, a small petrol engine that one hand-cranked, then used to fire the primary motor. There was no other way of starting *Grandma*. Like many small, two-stroke petrol engines of great vintage, the pony motor was notoriously unreliable, and often required extensive cajoling, stripping of carburettor, servicing the coil, etc, but on this occasion nothing worked. To make matters worse, the crank handle was too close to the mainmast, so that it was easy, especially when exasperated, to skin one's knuckles. Kurt was soon soaked by perspiration, stinking of petrol and bleeding. Jim and I both tried our luck but the pony-motor continued to sulk.

We drifted for three days. It was sunny and warm during the day, with a glorious, flat sea, though I found it hard to enjoy the weather. The nights came early and were cold.

One night, when Jim called me up for my watch at 2200, there was a single red light visible to seaward of the schooner. 'It's been there for some time, on exactly the same bearing,' he said.

'That means it's on a collision course,' I replied, remembering the incident on *Ice Bird* in 1976.

'It doesn't seem to be getting any closer,' Jim said. 'Maybe it's also a sailing boat that's becalmed.' Since there were no other lights visible, this seemed plausible, so Jim retired below. I continued to watch the light.

I decided it was getting brighter, although it is sometimes hard to be sure at sea, without reference points. Then I began to hear the occasional rumble of engines. I called Jim back up. By this time, the throb of motors was distinctly audible. *Ishmael* had no navigation lights, not even oil lamps, though we used them below. The only electric devices aboard were a couple of small dry-cell torches. I tried waving one in the direction of the oncoming vessel but it seemed hopelessly feeble.

Jim stuck his head into the deckhouse and called Kurt. He exhibited none of my nervousness with Kurt, and Kurt, in return, did not respond to him with the irritation he showed me. Kurt brought out some pieces of red and green plastic which he put over the torches, but this, in my opinion, only made them less visible. I was unwise, perhaps, to say so.

'So, what do you suggest?' Kurt growled.

By this time, the other vessel was clearly visible. It was a trawler of about 60', and now we could see both its red and green lights; it looked like it was going to hit us amidships. Its aft decks were floodlit, and we could see people working the nets. High-decibel rock music saturated the night, and from the way the bows swung a little to port, then a little to starboard and back again, always returning to a collision course, it appeared that the vessel was on autopilot, with nobody in the wheel-house. I wondered if I would be able to leap aboard the trawler as it smashed into *Ishmael*, or whether I was destined to end up swimming in the slimy, black sea.

I rushed below and grabbed two big pots. Returning to the deck, I began banging them together like cymbals, and Jim and I yelled. The cacophony sounded like a demented chorus. Kurt remained silent, though he looked pensive. He shone one torch at the trawler, shielded in red plastic, and with the other, illuminated *Ishmael's* American flag. Just when collision seemed imminent, the trawler's motors suddenly backed off. It continued to glide towards us, but a gentle breeze chose that moment to waft across *Ishmael's* deck, and the schooner eased clear.

'So, what were you worried about?' said Kurt, but I noticed that he put a very large dash of rum into his next coffee. The rum was a present from me, even though it severely dented the funds I'd come aboard with. Kurt, who was completely out of funds, had lamented several times that it was going to be a long trip without rum, so I bought him a case. It had been well-received, though hadn't thawed relations between us as much as hoped. In hindsight, it might have been better to sober him up.

In the inexplicable way of small, two-stroke petrol engines, the pony-motor condescended to start the next morning. After a futile attempt to breast the foul tide at French Pass, we anchored off to wait for slack water.

*Anchored just south of French Pass. Note the mainboom jaws, held to the mast with a wire parrel, rather than a metal gooseneck. This set-up was to cause us some grief in the not too distant future.*

It was beautifully warm during the day, after the sun cleared the surrounding hills at about 1000, but the temperature plummeted immediately when shadows returned after 1400. The nights were bitingly cold, and I wondered what it was going to be like in the Southern Ocean as autumn progressed. Combined with the stress I was feeling about my relationship with Kurt, I was beginning to slide, yet again, into depression.

The following day, we motored through French Pass at slack water without difficulty. Around the next headland, we came up to the township of the same name, which was just a few, single-story buildings nestled beside a public wharf, surrounded by lush green hills dotted with grazing sheep. The calm, dark-green waters of Marlborough Sound mirrored the surrounding hills, while the cloudless sky above was the deepest blue I had ever seen.

Kurt ran *Ishmael* ashore next to the public wharf. As the schooner struck the beach, the doors of a large building nearby burst open and dozens of screaming children ran out. They thought New Zealand was finally being invaded by the pirates of their dreams and story-books, and did not intend to miss one minute of the action. Their genial schoolteacher ambled out behind them, grinning.

After the schoolchildren, came old Tom Webber, the ruddy-cheeked, white-haired storekeeper and postmaster. Softly-spoken and deliberate, he'd been born in this little community 70 years previously and had lived there ever since. He only left once, briefly, during the Second World War, and then only to nearby Blenheim. His father had been born at this little outpost too, the first white baby in the region.

*It is approaching 1400, and already shadows are beginning to lengthen.*

It took some time to convince Tom that we had run the schooner aground intentionally. He'd already called for assistance on his radio, and shortly afterwards a trawler steamed into the anchorage. Luckily, its crew were happy enough to spend Saturday night in town, if you could describe French Pass so grandly. We had a great party on *Ishmael* that night. The trawler's crew amused me, consisting of three large-boned, fierce-looking, tattooed Maoris, and their cabin boy, a skinny, blonde-haired, pale-skinned youth with earrings, to whom they showed great affection.

Later that afternoon, Jim and I tried to buy some bread from Tom at the store, but were told, very apologetically, that everybody around there baked their own. In the evening, he brought a tray of savoury scones down to *Ishmael*, saying, 'My wife is concerned that you have no bread.' Of course, we had plenty of biscuits aboard, but there's nothing like fresh bread. The scones were delicious.

It took a week on the beach to complete the work on *Ishmael*. The biggest job we had to do was jury-rigging the delaminating centreboards. At low tide, we dug holes in the sand to pull the boards out, then strapped them up with fencing wire and shoved them back in.

Throughout this period, Kurt was in an extremely foul mood. The more he shouted at me, the more jittery I became, prone to fumble, drop things and misunderstand. I began to feel unnerved.

The thing that unnerved me most was an incident with the bobstay, which was connected by a shackle to a large eyebolt at the waterline. On the way down to Sydney from Fiji the previous year, this shackle had broken. This did not immediately bring the rig down, as it might have done on a modern yacht, though it needed to be repaired before the ship could carry sail again. To replace the shackle, Kurt had jumped overboard with a rope tied around his waist, while a nervous Ginna clung to the other end. It took hours.

'Damned near drowned me,' Kurt had laughed, when recounting the story over a few rums in Sydney.

On the beach at French Pass, I discovered that this jury-rigged shackle, far too small for its application but the largest available at the time, had never been replaced. It had elongated, and its pin was lying at an alarming angle. It was obviously about to fail, so I drew Kurt's attention to it.

'Yeah, yeah,' he said, without much interest, turning back to another task.

'Shall I replace it?' I asked, since I wasn't doing anything particular at that moment.

'No, give me a hand here.'

The days went by and the shackle remained on the job list. I reminded Kurt of it once in a while but was always rebuffed with growing irritation. It became an obsession with me, however. I had visions of being sent over the side on the end of a rope in the Southern Ocean, or of the rig coming down. I also began to look at the rest of the gear, and to wonder, probably unfairly, what *else* was about to fail.

Inevitably, there was a showdown. Kurt sighed. 'Oh, all right, damn it. How the hell are we going to have any adventure on this trip with you on board?' I could not fathom the look on his face, but the shackle got changed.

We pulled the schooner off the beach on 19 April, with the aid of our friendly trawler, since we'd missed the top of the tide and were almost stranded for another month. It took another two days to free the forward centreboard, which was now jammed in its case.

Kurt was despondent. 'Sometimes,' he said, 'I feel as if I am aground on my own garbage.' When he was in this sort of mood, revealing his vulnerable side, I loved him.

# Chapter Sixteen

# Sailing to Tahiti

We sailed on 22 April. As a farewell gift, old Tom and his wife presented us with a huge, cooked lamb roast, and a box of fresh vegetables from their garden. *Scurvy medicine*, they called it. After hoisting sail, we made a final pass in front of the wharf, where the Webbers, the trawler crew and others, were gathered to wave us on our way. *Ishmael* turned its bows towards Cook Strait and the Pacific Ocean.

There was little wind initially and so we motored, trying to get clear of the Strait by nightfall, since we had no chart of this area, or of any other, for that matter. We had a couple of passage-planning charts, showing wind and current, that spanned the entire Pacific Ocean, and that was it, besides a few out-of-date pilot books. For daily plotting, I had the Baker position line charts I'd used on *Ice Bird* in 1976. It was enough to get us across the ocean, but approaches to land would need caution.

After several hours, it became obvious that we would not clear Cook Strait before dark. Without a detailed chart, we didn't even know what course to steer once the night engulfed us. All we could do was steer east and hope that a shoal or promontory did not lie between us and open water. The pilot book indicated clear water ahead, but my fear was that we'd stray closer to the South Island.

We didn't even have a decent compass. The main compass, in its binnacle on deck, had a huge bubble in it and was unreliable, while the only other compass aboard was one of those boy-scout things with a folding lid, that usually sat on the gimballed table in the deckhouse.

A freshening NW breeze sprang up, the motor was shut down, and *Ishmael* swooped into the void. My watch was from 2200 until 0200. The ship was steering itself with the wheel lashed, the aft centreboard all the way down and the forward board up, but I stood by the wheel, anxiously staring ahead. There were no lights visible, no stars overhead, and the wind seemed to pass right through my inadequate clothes. Behind me, in its kennel of old car tyres lashed to the taffrail, shielded from wind, rain and spray by a plastic mackintosh, the ship's dog whimpered.

I was relieved to get off the deck at the end of my watch, but an hour later Kurt called me back up. A lighthouse blinked off our port beam. 'Do you think it's on North or South Island?' he asked.

If it was on North Island, we were drawing clear. If it was on South Island, however, we were sailing into trouble. With a rising westerly wind at our backs, we could well find ourselves embayed, and the schooner was not at its best against the wind. After reading the pilot book, I decided it had to be North Island, even though the sequence of the light bore no resemblance to anything I could find.

'You'd better be right,' Kurt growled, as the ship plunged on.

It was totally overcast now, with dark clouds scudding across the sky, and the smell of rain in the air. Although it was my watch below, I stayed on deck. I couldn't have slept anyway, as my stomach was knotted with anxiety. The ship was really straining in the rising wind, bucking and heaving, throwing sheets of spray across the deck, and beginning to round up in the heavier gusts. *Ishmael* would not self-steer properly when over-pressed, so I took the helm to hold a steadier course. Eventually we put the light behind us.

'I think we'd better get the main down,' Kurt said. He'd been reluctant to lower it sooner, in case we had to turn around suddenly and claw our way out of trouble. 'When I let the halyards go,' he added, 'grab the mainsheet and pull the boom into the gallows. You gotta be quick, or the boom will fall into the sea.'

This was because *Ishmael's* lazyjacks, while effective at containing the lowered sail when the boom was in the gallows, were not strong enough to take the weight of the boom unsupported. I looked fearfully at the boom. It was 36' long and 6" in diameter, solid wood, like a bloody great tree trunk. The thing that worried me most was hauling in the sheet. It was cleated to the leeward rail, meaning I'd have to leave the wheel and slither down there, get the sheet off the cleat and haul like crazy. What the schooner would do without someone at the helm, I was unsure. It was taking a lot of effort to hold the ship on course.

'What about getting Jim up,' I asked hopefully.

'What's the matter with you?' Kurt growled. 'I sailed this boat all the way from Honolulu to Sydney with just a girl aboard. Are you saying you're not as good as some girl? Give the wheel an extra spin, then let it go and grab the sheet.'

Ginna, in my opinion, was far tougher than I'd ever be, but I said nothing. Kurt went forward to the mast and cast off the halyards. As soon as I saw the halyards come off their belaying pins, I gave the wheel an extra couple of turns and slid down the canting deck to the sheet. Grabbing it, I heaved, but it remained taut. I couldn't even get it off the cleat. It was stretching like an elastic band and I couldn't gain an inch. *Ishmael* was beginning to broach. Another minute and the ship would be broadside to the wind and waves, with all the sails flogging. I looked anxiously back. The gaff was still up aloft. Kurt was heaving on the downhaul, to no effect.

I scrambled back up the deck to the wheel, fighting to get the ship back on course. Kurt was shouting something, but I couldn't hear what he was saying over the sound of the wind. Suddenly the gaff came loose from whatever it was hung up on and began coming down fast. I spun the wheel again and leapt for the sheet, but it was too late. The end of the boom was already in the sea. An extra-steep wave chose this moment to slam into the stern, and. with nobody at the helm, the ship slewed around beam-on.

The whole boom was overboard now, along with half the mainsail. *Ishmael* had no metal gooseneck on the front of the boom, just a set of jaws sitting on a timber flange screwed to the mast, and the boom was held to the mast with a wire strop, or parrel, around the front of the mast. This wire strop had broken under the load, the boom jaws came off the mast, and the luff lacing on the sail had also broken. The spar was now smashing into the hull every time *Ishmael* rolled, and Jim appeared, startled, in the hatchway. With the three of us heaving, we managed, after an hour or so, to drag the boom and sail back on deck and tie it down.

Relieved of the mainsail's pressure, *Ishmael* was now happily self-steering, and continued to do so for several days. The day following this incident, the sea was choppy, and a deep, gloomy green, beneath greasy-looking, grey clouds. After a momentous struggle, and much swearing, the boom was re-shipped, but we did not hoist the sail. From then on, my anxiety rose and dropped in concert with that bloody sail.

On the 25$^{th}$, *Ishmael* crossed the date line, so we had two Fridays. I knew all about the theory, but it still seemed weird, almost supernatural. From that day onward, I felt some uncertainty about the date, something that had never happened on the *Ice Bird* voyage.

Then, that night, when Jim woke me for my watch, he said, 'I know it sounds stupid, but I think something is following us.'

I laughed, but not as cynically as I might have done elsewhere. Out there in the Southern Ocean, in that immensity of empty, desolate sea, the sureties of urban existence easily crumbled. On *Ice Bird*, the Coral Sea had seemed a bit spooky, a little remote occasionally, especially on dark, windy nights, but that was nothing compared to this.

To give you an instance of what I am talking about, black-browed mollymawks, a sort of albatross, often flew beside the ship, always on the windward side. Hovering just above head height, they stared down at the person on the wheel, intently watching every move. The wind would blow them past us eventually, then they'd circle around and take up station again, always to windward. An old sailor's superstition claims that these birds are inhabited by the souls of drowned sailors, and it was not so easy for me to dismiss this notion as it might have been ashore.

*The thing* was clearly visible, about 30-40m astern. As each wave rose, the silhouette of a large, round shape reared up, like the head of a whale, only to sink back into the succeeding trough, but it never came any closer. The wind was squally, and the schooner was speeding up and slowing down as each gust passed. The thing kept pace exactly.

I became suspicious. 'It's like we're towing something,' I said.

'But we're not,' said Jim, 'I've checked. Look. He switched on a torch, shining it down into the water. There was no line visible. Then he shone it aft. Something pink gleamed briefly.

'Careful,' I said, 'you might aggravate it.' I'd heard that crocodile eyes glow red in the dark. Maybe the eyes of whales or giant squid do too.

There was nothing further that Jim could do, so he retired to his bunk. Besides, it was bloody freezing on deck. I retreated into the wheel-house, peering out occasionally, except when I had to steer in squalls. I felt nervous on deck, as if the thing might reach out and snatch me into the slimy deep. I'd read terrifying accounts of ships being attacked by giant squid, slimy tentacles wrapping around the masts like elephant trunks. As a child, I'd seen a 30m Blue Whale carcass that had been preserved and carted around country fairs. Poor *Ishmael* wouldn't stand a chance if attacked by something like that. It would be *Moby Dick* all over again. Sometimes I regretted that I was so captivated by stories.

I was looking forward to handing over the watch to Kurt, though I never looked forward to waking him. For one thing, I always felt uncomfortable going into the after cabin, which was his domain. The only other time I went there was to get the sextant. The major problem, though, was that Kurt was notoriously difficult to wake. He resisted, and you had to shake him so hard it was almost like assaulting him. Then he'd come up swinging, as if you were. The trick was to make him a cup of coffee beforehand, thrust it under his nose, and then beat a hasty retreat. The more rum you put in it the better.

He would be up in a flash, however, if you drove the boat too hard in squalls. That had happened several times on my watch, particularly at night. I was reluctant to crawl out to the tip of that long bowsprit while the others were sleeping below, to pull down the flying jib, even though I understood the risk of blowing out that lightly-built sail, or, worse still, of bringing down the fore-topmast. It was not the accepted thing on *Ishmael* to ask someone else to come up and stand by you, so I'd procrastinate, and then feel ashamed when Kurt stomped up on deck, cursing, and did the job himself. As if to shame me further, he'd just run out along the spar, then drop down and straddle it near the tip of the jib-boom.

He was openly sceptical when I told him about *the thing*. He wouldn't admit he could see it, even when I flashed the torch and its pink eyes flashed back. I noticed, however, that he was quick to return to the deckhouse. Later, when I was lying awake in my bunk, the clink of his rum bottle sounded more frequently than usual. I drifted into an uneasy sleep.

Kurt's bellow woke me just after dawn. 'Come up here and see your bloody whale.'

Jim was already on deck. Skipping along behind the schooner was a large, pink, plastic balloon, about 2m in diameter. A 200m ball of high-quality rope was tangled in the bobstay, the line then running back under the hull, where it had snagged on the fully-lowered aft centreboard, and then to the balloon. We'd been unable to see the line because it was about 3m below the surface where it emerged behind the boat. After a struggle, we salvaged the rope and lashed the pink whale on deck.

The following week brought miserable weather. It was not really a gale, quite reasonable, I am sure, for these latitudes in autumn, but conditions aboard became bleak. The wind was squally, and the chaotic seas frequently broke aboard. We lived in oilskins and boots. There had always been a leak in the hull - Kurt suspected the forward centreboard trunk - and we had to pump the bilges dry every watch if possible, or it would slosh over the cabin sole in the forward cabin, aka the cargo hold, where Jim and I slept, making the cabin wetter than ever.

It was not always easy to man the pump, which was mounted on the welldeck just forward of the deckhouse, on the starboard side of the mainmast. The pump was a very old type that looked like a toilet bowl and required priming with a bucket of water to make it work. Visitors in port sometimes mistook the pump for a toilet, especially since *Ishmael* didn't have a crapper below.

You had to fill a bucket of water from over the side, making sure the bucket did not drag you in with it, then pour the water into the pump, belay the bucket lanyard so you didn't lose the bucket to the next wave that swept the deck, snatch the long metal handle out of its bracket on the forward wheelhouse bulkhead, insert it and heave.

If you were not quick enough, the priming water would drain into the bilge, requiring you to repeat the process, or sometimes the boat would lurch wildly and fling the water back out of the pump, or you'd miss and throw the water onto the deck. If you were lucky, a wave would break over the welldeck at the right moment and prime the pump for you. My greatest fear was that I'd lose the irreplaceable pump handle over the side in one of *Ishmael's* violent lurches.

To make matters worse in the forward cabin, there was also a deck leak at the mainmast partners. Several times an hour, a wave would break against the topsides and flood the welldeck, causing a stream of water to flow down the mast into the bilges. Everything was wet and slimy, and stank of mildew. It was impossible to get warm; there were just degrees of coldness. The only thing that was worse than trying to sleep in that cabin was when you had to go outside and steer during squalls. The raindrops were so cold and dense they stung your skin like a horde of marsh flies.

Trying to sleep, as the boat crashed and banged along, as bilge water sloshed back and forth across the internal ballast in the shallow bilges, and as seas sluiced the decks, while huddled under damp bedding in stinking clothes, was almost impossible. When I did doze, I would drift into exhausted nightmares. When I awoke, I felt bruised and tired. On the day that bilge water got into my boots, which had fallen over on their side on the cabin sole while I slept, wetting my last pair of dry socks, I wept.

Kurt seemed to be equally run down, but Jim, being a South Islander perhaps, with good, Scottish blood in his veins, was more used to this climate. Besides, he was not the sort to complain, and even seemed to be enjoying himself. He continued to cook interesting meals, as well as stand his watches. Sometimes he sang.

Despite the weather, I spent quite a bit of time on deck taking photographs. Keeping my camera dry was a challenge, but I'd had a lot of practice looking after the sextant in these conditions, and had developed a bit of cunning. I'd also developed my skills aboard *Mistral* the previous year, getting some pleasing results. Despite finding this passage so challenging, I knew I was getting some excellent images.

When the weather was rough like this, and we were close to exhaustion, we did not always gather in the deckhouse for meals. We would fix our own breakfast and lunch, snacking on whatever we could lay our hands on. Jim would then cook our main meal during his evening watch, from 1800 to 2200, and eat his share. I would eat during my watch from 2200 until 0200, and Kurt would finish the food off when he came on watch at 0200. We ate straight from the pot.

The problem with this arrangement, for me, was that I was often still hungry after I'd eaten half of what was left in the pot. Jim, being Jim, would have been scrupulous in only taking his share, but I always struggled with temptation. Just one spoonful more... and Kurt's portion got smaller and smaller.

One night, while I was struggling with my demons, I heard my conscience say, quite clearly, *I must not eat my father's food.* I corrected myself, saying, *Kurt, Kurt's food*, then realised I could not separate the image of Kurt's face from that of my father in my mind. It was weird. I knew that Kurt was lying asleep in the aft cabin, and my father was in South Africa, but I could not recall, as I gazed longingly at the food, just who was who.

Shortly after this, we lost most of our dishes. I usually washed them, though there was no formal agreement, as there was about navigating and cooking. Sometimes, if I was tired or depressed, Jim did them, though it seemed unfair on top of the cooking, which took a great deal more time and effort than navigating out here in the open ocean. Coastal navigation is another game. On this occasion we were both exhausted and we let the dishes pile up in the large plastic drum we used as a sink. During Kurt's graveyard shift, from 0200 to 0600, he heaved the overflowing bucket out of the deckhouse onto the deck, to remind us perhaps. I presume that schooner captains don't stoop to washing dishes. Unfortunately, *Ishmael* was flinging itself about with abandon, and must have flung the bucket right over the side.

Then the large gas bottle ran out. Kurt swore he'd filled it before leaving. It should have lasted months. The bastards had cheated him. The one small bottle remaining would not last all the way to Seattle, or even Honolulu. 'We'll have to stop somewhere in French Polynesia,' he said.

French Polynesia, Tahiti, the Society Islands... I pricked up my ears. Suddenly, I began to feel a little bit of the old excitement. There had been precious little of it so far; the sense of confinement and desolation had been almost unbearable. I'd *seriously* begun to consider giving up sailing, even though the idea shamed me.

During all this time we'd been driving ESE, slowly getting deeper into the Southern Ocean. Already, we were at 44°S. The sun, when I took my noon sextant sight, if I could get one, was a mere 35° above the horizon. It seemed crazy, to me, to go further south in autumn. It was just asking for a hiding.

Kurt disagreed. 'The further south we go, the shorter the distance,' he said. That would have been true if we were following our original plan of sailing east until we picked up the Humboldt Current off South America, then swinging up past the Galapagos Islands and on to Hawaii.

'Not if we're going to Tahiti,' I retorted (bloody bolshie crew).

'We'll end up going backwards,' he replied.

'No, I think we can hold a NE course,' I persisted. 'Anyway, I think the wind is going to swing to the south. Look at the swell.' There was a deep, ominous swell rolling up from the south.

My real motivation was an obsession with getting north. I was a bit like Fritz on *Solo*. Kurt would have been well aware of the swell, as he was a damned sight better seaman than I would ever be. He went below without saying anything, and I continued to complain loudly, standing on the aft deck above his bunk. He stomped back up. With his scruffy clothes, gammy leg, unkempt beard and blood-shot eyes, I thought for one mad moment that he really *was* Captain Ahab.

'You'd better be right,' he growled. Luckily, I was. We struck the mainsail, putting the boom in its gallows, gybed the ship, and *Ishmael* raced away to the NE. By dusk, the southerly swell was enormous. We took in the jib while there was still enough light to see, which made working on the bowsprit safer, leaving the schooner under foresail only.

Just after dark, the wind switched to the south and began to blow with storm force. After a struggle, we lowered the foresail. The gaff appeared to hang up on something aloft. The centre deck was awash by this time with breaking crests, which did not make the task any easier. *One hand for yourself, one hand for the ship*, is the old sailor's saw, but on a night like this you need to be a bloody octopus.

*Ishmael* continued running at speed under bare poles. It took a lot of effort on the wheel to keep the ship running true. A small storm jib sheeted flat fore and aft may have helped, but we didn't have one. The seas rapidly built up on the existing swell. It is hard to tell true wave height from the deck of a small boat, but I think they were 10m or more. Some waves, which came in sets of three or more, were substantially larger, almost twice the size.

These larger waves were always accompanied by, or followed soon after, by squalls of stronger wind. The wind in these squalls was peculiar. It came in bullets, as if fired from a machine gun. It's hard to know how strong the wind was; it had a similar intensity to the storm on *Poeme*, where the wind strength was recorded independently at 70 knots, but the seas were incomparably worse, especially the larger sets.

One of these larger waves picked *Ishmael* up like a bath toy, a 32-ton bath toy, and hurtled the ship forward. *Ishmael* was moving so fast that *Grandma* turned over, in spite of the variable pitch propeller being set to the finest possible setting (it was damaged and could not be set to neutral).

The bows plunged into the trough of the sea ahead, burying the bowsprit and the foredeck right up to the foremast. At the helm, my feet came out from beneath me, and I began falling down the deck feet-first, saved only by clinging to the wheel. Then *Ishmael's* buoyant bows reared up, sweeping tons of water aft. The ship gave a violent roll to starboard, then to port and back again, like a horse shying away from danger. The centre deck was awash in green water for several moments, before the next wave came along and flung it all back where it belonged.

Good one, *Ishmael*. It is impossible not to love a ship, and your shipmates, when you have been through something like this together. Kurt had conceived and built this schooner with his own hands, and at that moment, I was in total awe of him. Just then, he stuck his head out of the deckhouse. 'What the fuck are you doing up here?' he asked.

The sound of *Grandma* turning over had nearly given him a heart attack, and he'd been flung out of his bunk onto the bulkhead. Oily bilgewater had sloshed up onto the deckhead in the forward cabin. It also got into the lockers beneath my bunk, which were over a metre above the cabin sole, destroying my camera, and all my rolls of film, among other possessions. I lost all the photos I had so painstakingly taken of *Ishmael* storming through the Southern Ocean. Kurt promised me he would send me copies of the photos he had taken, but he never did.

It was so cold on deck during the storm that we only took short turns at the wheel. Even then, our hands and feet ached unbearably, as if they were being squeezed in a vice. Of course, none of us had suitable clothing.

Time became seriously warped. After 10 minutes on deck, it seemed like I'd been out there for hours. I even accused poor, patient, Jim of abandoning me out there, only to be told that *I'd* hardly spent any time on the wheel at all. I don't know where the truth lies. It was the closest that Jim ever came to calling me a whingeing bastard.

All that night, and most of the next day, we ran under bare poles. The wind moderated slowly throughout the day, the sun came out, and, with great difficulty, I took sextant sights that showed we'd made good 120 miles without a scrap of sail up.

Taking sights in these conditions requires both skill and patience. You have to wedge yourself in position, in *Ishmael's* case leaning against the windward side of the deckhouse, feet braced against the bulwark, and wait for the ship to rise to the crest of a wave. Then you whip the sextant out from under your waterproof jacket, where you have been sheltering it from spray, and try to bring the sun down onto the horizon. You have to be quick, not just because the next wave that comes along might soak the sextant and risk tarnishing the mirrors, but because a wave might get between you and the horizon, resulting in a false reading of the sun's altitude.

On that occasion, I was on deck for more than 20 minutes to get a morning position-line sight, and again for the noon sight. In the sea running, it was difficult to get a good horizon. Actually, I missed the noon sight, because the sun hid behind a cloud at the critical moment, but a position line shortly after ran almost east-west. Some navigators used to plot a series of sights on graph paper to assess their accuracy, but I belonged in the camp that liked to focus on getting one good sight. After a while you get a feeling for it.

The sea looked tremendous, as if we were sailing up and down the hills of New Zealand, except that they were vivid blue, with large swathes of foam glistening in the sun like patches of snow. There was no danger now, the waves were evenly spaced, with a long fetch, and *Ishmael* rode them easily.

For the next few days, we ran fast before fresh SW winds. The decks were awash with breaking crests and the motion quite violent, but it seemed tame after the storm. *Ishmael* self-steered with the wheel lashed, the aft centreboard all the way down, and we took things easy. If it was necessary to correct the course a little, we adjusted the forward centreboard. A little bit up made *Ishmael* fall off the wind, a little bit down brought the ship's head up.

My most enduring memory of the great storm is not the wave that nearly pitch-poled us, but of being spoon-fed pumpkin fritters by Jim the next morning while I steered before the awe-inspiring seas. It was an act of such tenderness that it almost made me cry.

Then we ran out of water. Kurt swore he'd filled the tanks in Nelson. We assumed the tanks had sprung a leak during the storm. From then on, we lived on rainwater; luckily, it rained a lot. To add to the gloom, I dropped the pickle jar during an unexpected lurch, and broke it. It was a very large jar, given to us by Jim's friends in Nelson, and a much-valued treat. Most of the pickles were salvaged, and the broken glass carefully separated, though we had to eat them quickly, since we could not replace the vinegar preservative. My popularity was at an all-time low.

The next mishap was the loss of almost all of our remaining pots, crockery and cutlery. Luckily for me, Jim was responsible this time. From then on, he cooked in an old Milo can, and we ate out of empty food tins, sharing the last, precious spoon.

The first time we did this, we all got a mild dose of food poisoning. It was bums over the taffrail big time for a day or two. At least the temperature was a lot milder these days. In the stormy weather down south, having a crap had been a traumatic affair. Luckily, or perhaps as a result, I had been constipated most of the time. How the others coped, I didn't know, since bowel movements are not a comfortable topic of conversation for honest, seafaring blokes. Now, however, there was no waiting for a secluded moment. We crowded the rail, cheek to cheek.

There was one other memorable incident during this time, when I came on deck for a pee during Kurt's watch in the early hours of the morning. The schooner was sailing along briskly, self-steering across a boisterous, moonlit sea, low clouds scudding fast across the moon, bow wave roaring defiantly, but Kurt was nowhere to be seen. *Perhaps he's taking a nap in his bunk,* I thought, *since everything is as steady as she goes,* but I didn't really believe that, as he was too good a seaman. As he often said, 'I don't carry navigation lights because I expect my bloody crew to keep a good lookout.'

I looked forward. *Nope, not there.* I cast a furtive look into the aft cabin. *Not there either.* I went back on deck with quickening pulse. If he was overboard, it would be curtains; we'd never find him in this rough sea. *Vale, Kurt.* I was about to call Jim when I heard something. Plunk, plunk, plunk, plunk. It was Kurt's ukulele, a relic from his old Tahiti days. I'd seen it in the aft cabin, strapped to the bulkhead alongside the sextant, but he always refused to play it when asked.

The sound was coming from somewhere forward. I edged along the sidedeck and looked around the deckhouse. There was Kurt, sitting in the welldeck, with his back to the mainmast, oblivious to the occasional sea that sluiced over the bulwark. His wild, matted hair and beard were blowing in the wind, and he clutched the ukulele to his chest, fingers plucking the strings.

He was singing softly, and I just caught a snatch of the words, *Jesus wants me for a sunbeam.* It is an old Sunday School Christian song, but I think it was more a reflection of limited repertoire than spiritual conviction. He looked so innocent, so vulnerable, alone there in the moonlight, with his old stamping ground, Tahiti, somewhere ahead of the bowsprit, and a wake full of memories. The last time he'd played that ukulele, I suspected, he'd been a happy man, surrounded by his wife and children. I sneaked away and went back to bed.

Soon we were passing through the Austral Islands, outliers of the Society Islands, 600 miles south of Tahiti. *Only a week or so to go.* When I tuned the battery-powered, short-wave transistor radio in the early hours of the morning, listening for time signals to check the chronometer, I sometimes picked up Tahitian radio stations, and snatches of Polynesian music invaded the cabin.

Then the wind died. *Ishmael* lurched violently in the leftover swells, which struck randomly, first on the bow, then on the stern, then breaking aboard amidships. The motion began to drive us all crazy, even stoic Jim. We did not attempt to start *Grandma* because there was not enough fuel on board to motor any significant distance, and we were saving the fuel for emergencies or landfall.

We left the mainsail up, sheeted in tight, trying to steady the ship, to no avail, and also because a light breeze sometimes sprang up briefly, as if to tease us. During one of these episodic breezes, I was trying to coax the ship along when a vicious sea under the stern slammed the rudder over, tearing the wheel from my hands.

As mentioned before, *Ishmael's* steering apparatus, like much else on the schooner, was of the old school. Chains led from the trailing edge of the rudder, through turning blocks, up through hawsepipes to the deck, then to a drum forward of the wheel. Kurt had warned us about letting the wheel get hard over. Now that it was, the steering chain jumped off one of the turning blocks at the waterline and the rudder jammed. Then the force of the sea slamming into the rudder snapped one of the bolts holding the steering pedestal to the deck. We were in serious trouble.

It was at moments like this that Kurt showed his mettle. Despite being twice the age of his crew, and having a damaged leg, he never hesitated. He tied one end of the mainsheet around his waist and jumped over the side. Every time the ship rose on a wave, he'd wrestle with the jammed chain, and every time the counter slammed back down, he'd take a deep breath and hang on as he was dragged underwater. I really thought he was going to drown.

Somehow, he completed the task without getting drowned or losing his fingers, and we hauled him back aboard. A spare bolt was found for the pedestal, and then we lashed the wheel amidships, took the mainsail down, which was chafing holes in itself on the lazyjacks, and retired below. For once, I was smart enough not to mouth off about the gear, and Kurt, in turn, did not criticise my stupidity.

For a whole week, we drifted beneath an ominously overcast sky, meaning that I could not fix our position. At night, in particular, we listened anxiously for the sound of breakers on the surrounding reefs. Kurt entertained us with stories of how he'd wrecked his previous schooner, *Sea Gypsy,* on the barrier reef at Raiatea one night.

When the wind returned, it was from the SE. Hooray, the trade winds. *Ishmael* bounded off to the north before a steady breeze. Although the wind grew fresh, and the sea once again broke boisterously aboard, the sun shone brightly, puffy little trade wind clouds filled the sky, and the temperature steadily rose. We were finally emerging from what felt to me like a dark, slimy cave. I could not remember what it felt like to be warm and relaxed.

During our passage through the Southern Ocean, and in particular during our storm, I often thought about David Lewis, wondering how he had survived aboard *Ice Bird* during his Antarctic adventure. Not only did he survive, but he went back again in 1978 aboard *Solo*. In the early 1980s, he returned twice more on *Dick Smith Explorer*, telling me once that the Antarctic was the most beautiful continent on earth. I preferred the idea of Tahiti.

Now that we had clear skies again, I brought up the sextant. It was the first opportunity in over a week to get a fix. To my dismay, I discovered that half the silvering on the horizon mirror had tarnished black. I'd always cleaned and polished the sextant carefully between shots in the Southern Ocean, but had forgotten to do so during the calm. The problem was that the sextant box was screwed to the bulkhead beneath the leaky skylight in Kurt's cabin. It was a stupid place to store it, something I had pointed out several times. It was even more stupid not to clean the sextant daily. I could still use it, luckily, but the mirrors would need re-silvering at the earliest opportunity.

On 20 May, *Ishmael* was 100 miles south of Tahiti. I predicted sighting land shortly after dawn the next day, on the port bow. That night I didn't get much sleep. The daily fixes had varied wildly from my dead-reckoning, probably because Kurt did not keep a written log, and I had no idea of the course and estimated distance run when I was not on deck.

Adding to the complication, the working compass, as I mentioned earlier, was one of those piddly Boy Scout things, a small, circular brass unit with a folding lid. It sat on the gimballed table in the deckhouse. Once, one of us put a tin of condensed milk next to it, and for several hours we sailed in the wrong direction before I noticed.

I was on deck at dawn the next day, naked, for reasons lost in the mists of time, probably because it was warm enough, and my clothes were all disgustingly filthy and damp. Jim had just relieved Kurt of the watch but it was no time for sleeping. The sky was a jumble of blue-black clouds, and we all stared ahead expectantly. My stomach felt as empty as the horizon. I couldn't see anything that resembled land.

'There,' Kurt said, pointing off the starboard bow. The cloud he was pointing to had a slightly firmer upper edge. I would never have picked it, but he'd made a great many landfalls.

It soon hardened into a clearly-defined ridge. Tahiti. *Land-ho.* Making landfall after a long passage across the ocean remains the pinnacle of excitement for me. Those magical hours when the land first rises up out of the sea are the only periods in my life when I have genuinely known surcease from longing.

All day we ran towards it. The island grew larger but never seemed to come any closer. Late in the afternoon, we saw another gaff-rigged schooner silhouetted against the northern horizon, apparently going around the other side of the island. It was the shortest distance to Papeete, the main town and capital of French Polynesia, but that was the leeward side. Kurt preferred to stay to windward, like the mollymawks.

Night fell, and a violent squall descended on us. It was from the north, however, so we were quite comfortable in the lee of the island, drifting quietly under bare poles. This short gale caused havoc in the port, and to a couple of yachts making their approaches from the east, one of which was towed in a couple of days later by the French Navy, having suffered serious rigging damage. For *Ishmael's* crew, hardened by 30 days in the Southern Ocean, it was merely inconvenient.

The next morning, we were becalmed 10 miles south of the island. The view was stunning but we were impatient. It was time to fire up *Grandma,* and Kurt went below to do his thing. 'Hooray,' I said, as the pony-motor started first go, a minor miracle, but my smile was short-lived. When the bronze handle for *Grandma's* variable-pitch propeller, mounted on the deckhouse roof, was cranked to the open position, the engine began vibrating and making weird noises. Blue smoke billowed from the engine room.

Kurt dashed back down below, the motor stopped, and *Ishmael* slowed down. He came back on deck. 'The motor just fell into the bilge,' he said with a smirk. 'We're a sailing ship now.' Although this proved to be something of an overstatement, as a welded strut had just broken on the engine beds, it was effectively true. The motor was cactus, another victim of the storm that had pushed *Ishmael* to the limit.

We drifted for two days. At dawn on 23 May, we struggled past Venus Point, where Captain Cook had observed the transit of Venus, and Papeete finally came into sight. It took several hours to slowly tack into the harbour. Because of the seas breaking across the barrier reef into the lagoon, a constant current runs out of the pass, regardless of whether the tide is ebbing or flooding. It was enough to knock us back time after time. To leeward of *Ishmael*, when we were on the port tack, a rusty wreck lay on the reef, reminding us of our fate should we misjudge.

As usual, I was on the wheel, because Kurt always said that helming was the only thing I was good at. Just when I began to despair of ever getting in, the wind shifted direction slightly and gusted for a few moments. It was enough for us to squeak through the pass into the lagoon, 32 days out from French Pass.

# Chapter Seventeen

# The Islands of Desire

I gazed in rapture at Papeete's famous quay, where many of my heroes had tied their little ships stern-to, anchors out over the bow, at some stage in their voyaging careers. Tahiti is the lodestar of many voyagers' dreams, even if they deny it later, with sufficient salt water under their keels to know better. Ever since the days of early explorers like Wallis, Bougainville and Cook, Tahiti has held a particular place in the imagination of Europeans.

The quay was jammed with yachts. There was no way *Ishmael* could squeeze in there, even if we weren't under sail. Just past the quay, there was a small beach where anchored yachts took stern lines to the trees, but this, too, was crowded.

'Better tack,' Curt said. 'We'll anchor out until we get sorted.'

I put the helm down and *Ishmael* came about with a rattle of canvas, but then suddenly became sluggish. I was trying to bring the bow closer to the wind, but there was nothing I could do.

'You useless prick,' Kurt yelled. 'I thought you were the hot-shot helmsman?' He seemed to forget that I had just spent several hours on the wheel, tacking and tacking again, hands smarting, eyes stinging from the sweat of concentration, to bring the ship safely through the pass.

With a gesture of disgust, he let go the anchor, and the chain rattled out in a cloud of rust. We later discovered that the forward centreboard had chosen this moment to fall out of its case, and was dangling by one cable below the keel. If it had happened when we were transiting the pass, *Ishmael* may have been wrecked.

A man approached us in a dinghy. 'Hi. Good to see some folks still doing it the traditional way,' he said. 'I've just sailed up from New Zealand on a gaff-rigged schooner myself. We saw you on the horizon a couple of days ago, but took the soft option and motored in.'

His vessel, *Peregrine*, was a 35', ferro-cement, Pinky-sterned schooner designed by Jay Benford, and less than half the displacement of *Ishmael*, but it had made a quicker, easier passage than us, never getting below 30°S. Whether *Ishmael* would have fared better by staying further north one couldn't say, but I still gave Kurt a meaningful glance.

Our new friend gave us a lift ashore, as our 18' Whitehall skiff, lashed upside down on the welldeck, took a couple of hours to prepare, and Kurt was in a hurry to get off the boat. Jim and I were, too, but it was Kurt's decision, and it turned out to be a foolish one.

As we wandered along the quay towards the immigration office, dazed and delighted by the sounds and scents of the land, several people stopped Kurt with affectionate declamations. It had been more than a decade since he was last here, but he was one of those people that everybody remembers. A French guy, who Kurt had last seen in Honolulu, shouted, 'Kurt, you old bastard; bad pennies always turn up. Where have you been? We heard you'd run off to Australia with a girl young enough to be your daughter.'

Kurt, who was an intensely shy man beneath his bluster, visibly blanched. While they were talking, I glanced around at the yachts nearby, and nearly swooned when I recognised Bernard Moitessier's *Joshua*. Gazing at the bright-red steel ketch was like stepping into my imagination. I'd read Moitessier almost as often as I'd read Guzzwell and Peter Pye. I looked at *Joshua's* breathtakingly beautiful lines and remembered his poetic stories. Just then, drawn by raised voices on the quay, Moitessier stuck his head out of *Joshua's* companionway, but when he saw me looking his way, he quickly retreated, like a shy tortoise.

There was a sistership to *Joshua* moored nearby, and I have seen a number of others over the years, but none of them are as beautiful, or perfectly proportioned, as the original. Perhaps the designer, Jean Knocker, altered the plans in some subtle way after *Joshua* was built. I was keen to meet Moitessier, naturally. We had a couple of mutual friends, including David Lewis, who'd given me a message for him.

'Tell Bernard,' David said, 'that he should take *Joshua* to the Antarctic. Tell him I said *Joshua* is a perfect boat for the ice, and it is the most beautiful continent on earth.'

*Joshua in Papeete, soon after Moitessier completed his long voyage. When I saw the boat in 1980, a month or so before Moitessier sailed to California, it had just left the Naval Dockyard and was looking very smart. Photo: Bobby Schenk, SV Thalassa.*

Immigration was a bit of a problem, as Kurt and I did not have enough cash to lodge the required bond. Of course, we had no visas either. Kurt declared the schooner unseaworthy, and with its non-functioning motor, suspect water tanks, damaged sextant and dangling centreboard, that was no lie. Kurt said a few Tahitian words to the immigration officer, who smiled and shrugged. He gave us 30 days to sort things out.

In the end, *Ishmael* stayed six months and starred in a movie at Raiatea, where the crew had all expenses paid at the local hotel, something Kurt, apparently, made good use of, especially at the bar. Then the ship departed for Honolulu with a fistful of dollars in the kitty. It was a classic Kurt story, freewheeling through life, always landing on his feet, despite the occasional stumble. There was something about the man, the schooners he built, and the adventures he embarked on, that people found irresistible.

After leaving the immigration office, we went for a walk along Pomare Boulevard, wincing at the indescribable din from the frenetic traffic that hurtled up and down the street. Where could everybody be in such a hurry to get to on this small island? Nonetheless, the streets were lined with trees, the sweet scent of flowers prevailed over exhaust fumes, fantastically jagged, jungle-clad mountain peaks dominated the sky, and in the distance, beyond the barrier reef, the outline of Moorea, even more ethereal, was etched against the horizon. A young Tahitian woman, in a bright pareu, tossed her long, black hair and gave me a sweet smile.

At the local bank, I changed my last A$100 into francs, while Jim, being a good boy, lodged his bond. Kurt then led us down a narrow alley near the market and disappeared into a crack in the wall. It was a bar that he'd frequented in the old days, a place where Tahitians gathered, away from tourist haunts. There were three large-hipped, heavy-breasted, Gauguin ladies sitting at the servery, their ample flesh loosely wrapped in flowing pareus, flowers in their hair. Kurt was soon chatting to them in Tahitian. The women kept looking at Jim and me and laughing, uninhibited by their lack of teeth.

'What are they saying?' I asked.

Kurt grinned. 'They want to take you home and fuck you.'

I must have looked shocked, or frightened, for the women shrieked with laughter and made suggestive gestures. They spoke enough English to get the gist of things. Casual sex in Polynesia, it seemed, was a lighthearted affair. Relationships, I soon discovered, were more complex.

After a few beers, I asked where the toilet was. Following directions, I walked behind some large crates, for the bar also seemed to be some sort of warehouse, and found myself in a walled-off section of the alley. One of my admirers was squatting over a grate in the floor, her pareu delicately hitched up. I apologised profusely and began backing out.

'No,' she commanded, 'you pee over there.'

She raised an imperious arm and pointed towards the wall, down which a trickle of water gurgled. Self-consciously, I obeyed, vaguely remembering something I'd read about French public toilets being communal, though I've since been told that's bollocks.

Several beers followed. Jim generously bought most of them, however my money was still disappearing at an alarming rate. Kurt had no money, but we were happy to treat him. He'd brought us here, safely across the stormy ocean on his magnificent schooner. I also hoped that it might wash away the tensions of the voyage. After all, we'd initially forged our bond in alcohol, during those heady days on the Hawkesbury River.

At some stage, a drunken, bare-chested French guy wandered in and began trying to caress some Tahitian men sitting at nearby tables. They pushed him away good-naturedly but he persisted. Suddenly, one of the Tahitians stood up and punched him in the mouth, causing him to fall onto his backside. Sitting on the floor with down-turned mouth, a trickle of blood running down his chin and tears on his cheeks, the poor guy looked like a sad clown. I laughed involuntarily, and then felt instantly ashamed.

For the Tahitian women it was no laughing matter. They leapt off their stools at the bar and accosted the aggressor, shouting loudly and slapping his face. He protested, but they just slapped him harder. Then they helped the French guy to his feet, tenderly wiped his face, sat him down and bought him a drink.

When they returned to the bar, one of them said, 'They should not hit him. They know he is a *mahu.*' This word translates as man-woman, or transgender male, a common social and cultural element of Polynesian society. I am not sure if this social position embraces what the West calls a gay man.

The Tahitian guy, emboldened perhaps by distance, shouted something at the women in Tahitian, and one of them yelled back. She laughed. 'He says he did not hit the guy because he is a *mahu*, but because he is French. I tell him, better behave yourself, boy.' She leaned forward. 'You see, in Tahiti, the man is boss of the sea, and the woman is boss of the land. When the boys are ashore, they do what we tell them, or else.' She gave a throaty laugh and made an obscene gesture with her forearm.

By this time, I was beginning to feel dizzy, and was worried I might pass out. 'I need something to eat,' I said faintly. Nobody noticed.

After we'd visited another bar, and after I'd repeated my plea for food several times, one of the women said, 'OK, we go eat, good tucker.'

*Wonderful,* I thought, *authentic Tahitian food.* I was looking forward to being received into a traditional Tahitian home, just like my heroes had been, all those years ago. Who said the old Tahiti no longer existed? People kept saying, '*Oh, you should have been here twenty years ago. It's all ruined now.*' Rubbish.

The women led us to a Chinese restaurant. Besides the disappointment, I thought nervously of the bill. We'd been joined by an entourage from the second bar, including a Tahitian man named Pedro, which I thought was an unusual name in Polynesia, and his sister, Marie.

The situation with Marie was interesting. At the second bar, I'd been talking to Pedro, who told me he was a fisherman, but we'd been repeatedly interrupted by a short, slim, elderly woman with a face like a prune, who looked part Tahitian, part Chinese. She was wearing tight jeans, a checked shirt, a Chicago Bulls baseball cap, and cowboy boots, with a packet of Lucky Strike cigarettes sticking out of her shirt pocket.

She kept leaping onto my lap, grinding her buttocks against my pelvis, and shouting in my ear, 'You come American schooner. You take me Los Angeles.'

'No, no, I'd rather stay here,' I said each time.

Eventually, Pedro reached over and swiped the old lady off my lap. 'Go away, Grandma,' he said.

She swore at him, but soon appeared on the stage, dancing gaily with the lead guitarist, cigarette dangling from the corner of her mouth. Perhaps she really was his grandma, hence Pedro.

Then Pedro turned to me. 'If you want to stay in Tahiti, you marry my sister and come fishing with me, OK?' He waved his arm and shouted, 'Marie, come here!'

Marie was about 30, I hazard a guess. She was only slightly plump, with an open, pleasant face, *and* she had most of her teeth. By the standards of the night so far, she was a beauty. She coolly appraised me, and then agreed to the marriage, on the condition that I understood that *she* was the boss ashore.

'When I say enough beer, you no more drink, OK?' Then she turned to the bar and shouted, in a sergeant-major's baritone, 'More beer.'

An hour later, at the restaurant, she was still shouting, and there was no sign of food. I looked up at the ceiling, trying to steady myself. I could see, let alone feel, the rotation of the earth. Suddenly, I decided that I couldn't take it anymore, stood up, pulled all that was left of my money out of my pocket, put it on the table, and staggered to the door.

Marie rushed out behind me. 'Where you go?' she demanded.

I broke into a run and swerved down a dark alley. Luckily, she didn't follow. I suspect it may have been the shortest betrothal in history. Kurt and Jim would have to fend for themselves.

I had no idea where I was, but luck delivered me to the waterfront, where I sat down with my back to a tree. Nearby, I could see the silhouette of *Joshua*. As I sat there, mind gaping from alcohol, all the voyaging legends came to parade before me, retelling the stories I'd read or been told, a nautical version of Chaucer's *Canterbury Tales*. *They have all been here*, I reminded myself, *tied up to this very quay*. *I am a voyager*, I thought drunkenly, *I am a voyager, I am a voyager*. But I still felt like a fraud.

I remembered something Barry Lewis said to me when I returned from sailing *Mistral* to Bowen. He had been impressed by that voyage, and he has a high standard of seamanship. *All you have to do now*, he said, *is bring that determination ashore*. I did not feel that

I had acquitted myself well on *Ishmael*, but nonetheless, I recognised that sorting out my life was going to be my next challenge. It was time to take control, instead of endlessly chasing after some lost dream, left behind on the dock at Durban all those years ago.

Kurt and Jim arrived an hour or so later. I didn't ask them how they'd extracted themselves. Our immediate problem was getting back to *Ishmael*, since our dinghy was still on board. It was late, the other yachts were all in darkness, and the oily harbour didn't look inviting for a swim. It was surprisingly cold, too. We walked up and down, trying to keep warm. On one of these peregrinations, I failed to come about when the others did, choosing to walk a little further on my own. When I came back to the beach about 15 minutes later, Kurt and Jim were nowhere to be seen, but then I noticed that there was a dinghy trailing off *Ishmael's* stern. They'd borrowed it without permission from one of the other yachts. No harm was done, ultimately, apart from the 'schoonermen' getting an unsavoury reputation.

I resigned myself to spending the night ashore and sat down under a tree again, knees drawn up to ward off the chill. I must have dozed, because I was startled awake by a soft hand on my shoulder. Kneeling beside me was a handsome Tahitian boy, in his late teens by the look of him. He was bare-chested, wearing nothing but a pareu wrapped around his slim waist, with a *Tiara Tahiti* blossom behind one ear, and he exuded a faint perfume.

'What are you doing here?' he asked. I told him my story and he frowned. 'You cannot stay here,' he said. 'Bad men are coming and they will hurt you.'

I could hear a noise further down the street that sounded like garbage cans being kicked. 'Who are they?' I asked.

'They are my friends. I met them in jail, but I don't want you to be hurt.'

'Why were you in jail?'

'For fighting with the police,' he said, with a note of pride in his voice. 'We hate the French government.' He took me by the hand and led me to a small area near the pirogues that were pulled up on the beach. There was an overhanging crop here where I would not be visible from the road. 'I will come and visit you on the schooner,' he promised, giving me a Gallic kiss on each cheek, 'but now I must go back to my friends or they will come looking for me.'

I huddled down against the cold, knees drawn up to my chin. Dawn came at last, light seeping almost reluctantly into the eastern sky, just as it had on so many long night watches at sea. Ridge by ridge, the sun's rays lit up the peaks of Moorea, the island that many visitors, including Eric and Susan Hiscock, have called the most beautiful in the world. Unfortunately, in a pattern consistent with so much of my life, I never made it across that narrow strip of water.

Jim, always the stalwart, rowed in to pick me up soon after, and I gratefully crawled into my bunk for a snooze. I kept a lookout for the boy for the rest of my stay, but never saw him again. I also kept a wary eye out for Marie. I didn't see her either, but that wasn't quite so disappointing.

Later that morning, we found a place for *Ishmael* in front of the beached pirogues, putting an anchor out over the bows and taking lines ashore to the trees, in the time-honoured Papeete tradition. We could look out over the bowsprit to the pass, through which voyaging yachts came and went each day, with Moorea shimmering in the distance.

Framed by *Ishmael's* traditional profile, it seemed a timeless vista, especially when pirogues were racing across the lagoon, as they often did, training for the soon to start Heiva celebrations, which climax on Bastille Day, 14 July. Once, to my amusement, this image was rudely shattered by an international passenger jet that came screaming across the harbour at low altitude. The airport was just up the road.

There was an interesting group of singlehanded sailors in port. Utz, a lean, weather-beaten, deeply-tanned old German, was a retired fisherman, sailing the 11m, clinker-planked, double-ended cutter, *Frauken of Bremen*.

*Utz on the bow of Frauken of Bremen later that year in Whangarei Town Basin, New Zealand.*

He wore a salty, home-made canvas hat, and exuded an air of accomplished contentment. *Frauken*, like *Moonraker*, was a converted fishing boat, and I was fascinated by it. To my eyes, it was more beautiful than any millionaire's gold-plater. Utz was on his second circumnavigation. This trip, he confided, was much easier than the first, because he'd fitted an Aries self-steering gear. Prior to that, he used twin jibs in the trade winds, which keep the boat balanced, or hand-steered. Solitude did not bother him, but he did mention the joy of coming into port and seeing a familiar face.

Perhaps that was why he kept close company with a couple of other solo sailors in Papeete. One of them, a short, plump, Belgian man, who reminded me of Johann Trauner on *Lei Lei Lassen* all those years ago, had come through the pass just ahead of *Ishmael* on an 8m steel yacht that excited my interest. He followed Utz around like a puppy, but I found it impossible to get a word out of him. When I spoke to him, he looked petrified. It only occurred to me later that perhaps he couldn't speak English. According to the *Caribbean Compass*, Utz eventually completed four solo circumnavigations, before settling down in Bequia, in Saint Vincent and the Grenadines, where he cruised locally in his last years. He died in January 2009, in Venezuela.

Chris was another solo sailor, an English guy in his mid-thirties, with a shaggy black beard. He was sailing a 29' timber sloop, which had a slight reverse-sheer and large, vulnerable-looking windows. Apart from a Hasler self-steering gear, it did not appear to be customised for offshore cruising. Like most yachts of that era, he hanked on his headsails, requiring a lot of foredeck work, and mentioned how exhausting, and occasionally frightening, it was in heavy weather.

He asked keen questions about the Austral Islands and was disappointed that we had not sighted them. He shuddered when I told him about the boisterous easterlies that we'd encountered just north of there, and the lively, two-metre beam seas that had washed *Ishmael's* decks.

'I'm thinking of going down there,' he said. 'Need to get off the beaten track. Sir Francis Chichester is my hero, but so far all I've managed to do is blow downwind along the trades.' He hung his head. 'Can't even cope with that most of the time.' He sold his yacht in Australia.

A lot of the voyagers were talking about an Australian singlehander called Alan, who was on the final leg of a 15-year circumnavigation. He had last been seen in the Tuamotu

Archipelago, and was due in Papeete any day. His steel yacht had been squashed by a large ship alongside a wharf in the Canary Islands, and was now 30cm narrower, but he claimed that it sailed better to windward afterwards. He wasn't going to fix it until he got home, even though the topsides were completely rusty, and the splits in the timber cabin sides were covered with plywood. He was very funny, they said, the life and soul of the party, but a real loner when it came down to it.

I was shocked when somebody eventually mentioned the yacht's name, *Youth*. I remembered the crew of wild guys that Alan Quigley had aboard when he arrived in Durban, with whom he'd stormed around the South Pacific and Indian Oceans. I also remembered him sailing from Durban on the immaculate *Youth*, after a two-year refit, with five crewmembers who'd paid handsomely to join the Quigley Club, and wondered what experiences had led him to solitude. Unfortunately, he did not arrive in Tahiti before I left.

One of the most fascinating boats in Papeete was the Baltic Trader, *Sofia*, a 123', three-masted gaff-topsail schooner, built from Scandinavian pine in Sweden and launched in 1921. In 1969, 20 young people seeking an alternative life each put in US$1500 to purchase and refurbish the vessel, which was by then in poor condition. It did, however, have a reliable diesel engine, a 30-year-old, two-cylinder, June-Munktell model, which came to be known as *Junie-baby*. The ship was run as a commune, with the crew members democratically electing their captain for passages, but sharing equal responsibility for the ship in port.

After refurbishing *Sofia* in Spain and exploring Europe, they sailed to the Caribbean, where many of the crew returned to the United States, leaving seven core owners, who ran the ship as a cargo vessel for some time. By 1975, aided by a revolving cast of paying crew, many of whom became part-owners of the boat, *Sofia* had made it to New Zealand, eventually completing a circumnavigation.

As in all communes, social relations on board were dynamic at times, though those people that persevered formed fierce bonds. Somehow, the old ship kept on re-funding itself and rolling along.

In 1978, *Sofia* left Boston for the Caribbean again, intending to attempt a second circumnavigation. In early 1980, the crew were briefly arrested in Panama, victims of the acrimonious transfer of power in the Canal Zone from the American Government to local authorities. They were arrested again in the Galapagos Islands and fined $12,000 for entering without visas, but charmed their way out of that one. It was agreed that if the female crew members would accompany the Ecuadorian officials to a ball, they'd be permitted 24 hours ashore, an outcome that was satisfactory to all. The officials behaved like gentlemen and the crew had a fun day ashore.

*Sofia under sail. Photo: Pamela Bitterman.*

They had a Guatemalan anteater on board as a pet, though it wasn't very friendly. Most crew got bitten at one stage or another. Once, when *Sofia* had been barrelling along in the trade winds between the Galapagos and Marquesas archipelagos, one of the crew, Bill, had been so outraged by the creature that he threw it into the sea. Then, in a fit of remorse, he jumped in after it. It took the rest of the crew quite some time to start the engine and turn the old ship around. They found Bill treading water, blood streaming down his face, with the anteater sitting on his head. They were lucky to be rescued, and even luckier that the blood didn't attract sharks.

There were also hordes of gigantic, Panamanian cockroaches aboard. Anteaters don't like eating them, apparently; a big lizard might have been more practical. The crew weren't too fazed by the cockies, though. They were an easy-going crowd, at least when it came to critters. They sometimes took exception to each other, but mostly co-existed remarkably well for such a large crew. At times, there were 16 people aboard.

Early one morning, before the day got too hot, Jim and I rowed *Ishmael's* Whitehall skiff up to Maeva Beach, beyond the airport, where *Sofia* was anchored. We climbed up onto the deck and found several crew sitting side by side on the thunderboxes that hung off the stern, pants around their ankles, performing their morning ablutions while drinking coffee and chatting. They unblinkingly included us in their conversation. Clothing was optional aboard *Sofia* when away from the public gaze, and social attitudes determinedly alternative. There was one long-term crewman, Mother Boots, who was openly gay, and they all loved him.

A couple of the women decided that Jim and I needed haircuts. I should have accepted, as my hair had become a long, straggly mess, but declined. Jim had a trim, but was bitten by the anteater while sitting in the 'barber's chair'. It sidled past, looking sweet and innocent, like a giant cat, and he made the mistake of trying to pet it. He should have known better because we'd met it before, when we first went aboard *Sofia* alongside the commercial wharf in town. We were invited aboard by Bill, the anteater's best friend, who jumped over the bulwark ahead of us. Unfortunately, he landed on the anteater's tail. It ran off with a squeal, then turned abruptly, charged back and bit him on the leg. I often wondered if he regretted rescuing it.

Having told Bill that my relationship with Kurt was strained, he suggested I join *Sofia*, but I lacked the funds. They offered me a passage for free if I would take on the job of cook, but that is not my best suite, and cooking for a crew of 16 sounded daunting. I am also not at my best in a crowd, but wonder if the alternative dynamic of *Sofia's* crew might have been just the thing I needed. *When the going gets weird, the weird get going.* And I suspect that the basic fare dished up from Sofia's galley, based on things like potatoes, dried beans and tinned tomatoes, would have been easy enough to master.

Unfortunately, *Sofia* sank off New Zealand in February 1982, with the loss of one crewmember and the anteater. The keel had begun to hog badly, and they attempted to repair it in Nelson by bolting on a new hardwood timber externally. This was obviously not sufficient, and the boat opened up in the first bad weather they encountered after leaving port, sinking quickly. One of the long-term crew we met in Tahiti, Pamela Bitterman, later wrote a book about *Sofia*, titled *Sailing to the Far Horizon*. She also crewed aboard *Ishmael* for a few weeks around French Polynesia's Leeward Islands after I left the ship. Incidentally, *Sofia's* crew befriended Rex and his girlfriend, Louise, on *Mistral* in Nelson, after Rex's solo crossing of the Tasman Sea.

Kurt was making it obvious he wanted me to leave *Ishmael*, and who could blame him? The atmosphere on board remained strained. When we tested the water tanks, no leaks were found. 'I'd like to know who was having showers while I was asleep,' he said, though he could hardly have been serious. We'd all stunk like hyenas, and anyway it was bloody freezing.

Perhaps, during the storm, *Ishmael* had flung the water out of the tanks through their breathers. I never took a close look at *Ishmael's* plumbing, so anything is possible. With all the other leaks, we would not have noticed. Either that, or Kurt didn't fill the tanks like he claimed to have done. Ditto the gas bottle. I remembered that dodgy shackle we'd argued about in French Pass, and thought that it looked like a disturbing pattern.

I remembered one day, deep in the Southern Ocean, when Kurt had mentioned that he felt particularly low. 'I have never felt like this before,' he said, and I'd wondered then if he was suicidal.

Eventually I called some friends in Australia, Olive and Phil, and borrowed the money to fly home. They even gave me a job working at their marina when I got back. At the Papeete telephone exchange, I had difficulty making myself understood to the local operator, and was charmed when Bernard Moitessier stepped forward to sort out the problem. Afterwards, we chatted briefly. I was going to mention my friendship with David Lewis, and pass on David's message, but when I expressed admiration to Moitessier about his writing and voyages, he giggled and turned away.

I saw him around town on many occasions after that, once eating a hamburger in a cafe with a furtive expression on his face, as if he felt that he might be judged for not climbing a coconut tree, or diving for a fish, whenever he felt hungry, but he never met my gaze.

I could not help comparing Bernard Moitessier to David Lewis, who was completely unfazed by public attention, happily chatting to anyone who approached him. For David, accepting, even enjoying, public recognition was part of the story. For Moitessier, apparently, it was not.

David did not consider himself particularly famous, or behave as if he was, whereas I suspect Moitessier was very conscious of being a celebrity. I think he felt hounded by visiting sailors after a decade in Papeete. I don't think I'd like strangers knocking on my hull all the time, either. David would have just given them a piece of sandpaper and put them to work.

One thing about Moitessier that fascinated me, and provided a perfect illustration of his desire to keep things simple, was his dress code. I only ever saw him in one of two outfits, consisting of a pair of short trousers matched with a short-sleeved shirt, which he swapped daily. He'd wear one pair while the other hung in the rigging, drying in the sun.

Kurt celebrated his 50$^{th}$ birthday in Papeete. We bought some cheap, dreadful, Algerian wine and invited nearby voyagers and a few locals to a party. An old Tahitian man, a regular from the crack-in-the-wall bar, played Kurt's ukulele. Beneath the awning between the masts, with Moorea shimmering on the horizon and pirogues racing up and down the lagoon, I had a brief glimpse of how things might have been.

Then, typically, I blew the moment. I turned to Kurt and said, 'I can't imagine what it feels like to be 50.' I really couldn't, but saying it is something else.

Kurt just laughed. 'I still feel like I'm 25,' he said. 'But that's not surprising, I guess. The psychologist who interviewed me in prison reckoned I had an emotional age of 15.' He smiled enigmatically.

'I'm no better,' I said, and for a moment we were united in remorse. I know now what he was talking about. At 50 I still felt like I was a boy, and even now, at 72, I don't always remember that I am not.

On my last evening in Papeete, we had drinks aboard *Sofia*, which was once again tied alongside the quay for the night. They came down every few weeks to top up with water and supplies. Sixteen people eat and drink a lot. My head began to spin again from too much alcohol, and I staggered out of the cabin, lying down on deck, looking up through the topmasts at the stars.

Kurt, thinking that I'd gone for a walk, said a whole bunch of unkind things about me; what a useless crew I was, and how hopeless I was in general, no less painful because they were true. Then he said, in a strangely tender, wistful tone, 'But, gee, I really like him.'

There was an incredulous chuckle from someone, but I understood. Besides a mutual appreciation for traditional boats, we recognised something in each other; a deep yearning, that endless quest for a nebulous beauty glimpsed in the mythic distance, which has driven sailors beyond the horizon since time immemorial. An intrinsic aspect of that yearning, however, is an inability to articulate it.

The time finally came to catch my taxi to the airport. Kurt came to the rail and we shook hands. I said, 'Don't forget who it was who tacked your boat through Papeete Pass,' and walked away, turning my back on the famous quay that I had dreamed of for most of my life, and all the fascinating boats tied stern-to the Papeete waterfront. For a few weeks, I had felt as close to my dream world as I had ever come. A light rain was falling, and a rich, sweet aroma, like vanilla, rose from the damp earth. The beautiful mountains were hidden from view under the veil of darkness, as were my tears.

See *Last Days of the Slocum Era, Volume Two,* for the rest of this memoir.

# Acknowledgements

I have avoided completing this page in both volumes for as long as possible, since it is impossible to acknowledge, or remember everybody. For those that I miss, may I say, thank you sincerely for your support. The following list is in no sort of order, certainly not an order of merit.

Thanks to **Robin Lee Graham**, for his inspiration and the photos of *Dove*, to **Clive Rouse** for teaching me (or trying to), the basics of sailing and seamanship, much advice, and for numerous photos of *Clipper*, to **Margaret Worthington** for several photos and artistic inspiration, to **David Matzenik** for a lifetime of good fellowship and permission to use photos from the *Kelasa* collection, to **Adrienne Matzenik**, even though she can no longer read these words, for her unwavering friendship and inspiration, to **Rags Weldon** for some great photos and yarns about the old days, to **Geoff Collins** for giving me the first roof over my head in Australia, as well as the loan that allowed me to buy *Poeme* and leave him looking for a new flatmate, to **Julia Hazel** for the photo of *Jeshan* under construction, and for her inspiration, to **John Murray,** of the trimaran, *Unbound*, who accepted me for who I was without criticism in Durban, and who persuaded me to make the best decision of my life, to migrate to Australia, to **Ros Dawes** and **Gordon Lewins**, who were the best shipmates aboard *Ice Bird* one could ever wish to have, to **Rex Byrne**, for a great adventure on *Mistral*, and a lifetime of friendship, to **Jim Ferris**, fellow crewman aboard *Ishmael*, who showed me many kindnesses, to **Laurie Williams**, my first adult friend in Durban, for your ongoing friendship, and the huge stack of *Practical Boat Owner* magazines that kept me enthralled, and to **Terry McCartney**, **Marc Baudouin**, **Scott Mitchinson**, **Tony Nelson**, and all the merry pranksters of the Hawkesbury River, who made my youthful adult days there so much more interesting than they might have been.

# About the author

Graham Cox is the author of *The Junk Rig Hall of Fame*. He was born in Durban, South Africa, in 1952. In those days, Durban was one of the great crossroads of the ocean voyaging world, and from the age of 14, he crossed paths with many extraordinary sailors, both famous and unknown, who passed through that port, including Dr David Lewis, on the catamaran, *Rehu Moana*, who became a lifelong friend, and Robin Lee Graham on *Dove*, who became his great inspiration. Some of the voyaging sailors he met there became lifelong friends.

He migrated to Sydney, Australia, in 1972, where he met up again with David Lewis and helped him prepare *Ice Bird* for the Antarctic, as well as helping Colin and Rosie Swale prepare their catamaran, *Annaliese,* for their daring Cape Horn voyage. Over the next 50 years, he sailed on a number of voyages, in a variety of craft, and closely observed developments in ocean voyaging, giving him unique insights that he brings to *Last Days of the Slocum Era*.

He has a degree in English from James Cook University, Townsville. Besides writing *The Junk Rig Hall of Fame*, he has published numerous articles in magazines. He still lives afloat and dreams of blue horizons.

For more interesting cruising boats and commentary from Graham Cox, visit:

**https://randomboats.blogspot.com**

Made in United States
North Haven, CT
27 May 2025